George M. Gould

The Meaning and the Method of Life

A search for religion in biology

George M. Gould

The Meaning and the Method of Life
A search for religion in biology

ISBN/EAN: 9783337036164

Printed in Europe, USA, Canada, Australia, Japan

Cover: Foto ©berggeist007 / pixelio.de

More available books at **www.hansebooks.com**

THE MEANING AND THE METH
OF LIFE

A SEARCH FOR RELIGION IN BIOLOGY

BY

GEORGE M. GOULD, A.M., M.D.

"FROM LIFE, THROUGH LIFE, TO LIFE."

Ὁ Λόγος Σάρξ ἐγένετο.

G. P. PUTNAM'S SONS

NEW YORK   LONDON
27 WEST TWENTY-THIRD STREET   24 BEDFORD STREET, STRAND

The Knickerbocker Press

1893

CONTENTS.

CHAPTER		PAGE
	INTRODUCTION	1
I.	PHYSICAL AND METAPHYSICAL	12
II.	PARTIAL TRUTHS	24
III.	INCARNATION	58
IV.	CYTOLOGY	77
V.	SENSATION	101
VI.	EVOLUTION	145
VII.	REPRODUCTION	157
VIII.	CONCERNING EVIL	176
IX.	JUSTIFICATION OF THE INCARNATION PROCESS	188
X.	FREEDOM	199
XI.	PERSONALITY	211
XII.	IMMORTALITY	229
XIII.	ETHICS	239
XIV.	BEAUTY	280
XV.	SLEEP, DREAMING, AND AWAKENING	284
	INDEX	293

THE MEANING AND THE METHOD OF LIFE.

INTRODUCTION.

THE ordinary railway laborer, in cutting through a wall of stratified rock that by volcanic action has been tilted or warped out of its original level, never dreams of the causes of the stratification or of its displacement. To him the quarrel of the "Neptunists" and the "Plutonists" has never been distantly suggested. He daily sees these peculiarities, but to his incurious mind no question of why or how has been aroused thereby. When the railway cut has been completed, thousands of people are rushed through it, but doubtless to very few, if the train were halted in front of the wall, would the significance of the tilted and twisted layers be clear. Of the millions drawn by the locomotive engine, few understand the mechanism by which they are so powerfully and rapidly pulled along. Of these few in a million, least of all probably the engineer, who can use the force so perfectly,—of these very few, perhaps one may understand the nature of heat and the action of heated water, and how the combined bombardment of billions of crowded atoms striking the piston-head batter it back and forth and transmute atomic vibration into molar motion. Thousands daily pass an optician's window and are not moved to inquire as to the cause of the turning of the vanes in a Crookes' radiometer. And yet all these mysteries upon, among, and by which we live, are clearly and easily explainable, if our minds but stopped to inquire; if we desired to know, the knowledge would be forthcoming. One of the saddest things in man is his willingness to be ignorant and incurious of the mystery that he is and by which he is surrounded. A very discouraging exaggeration of this willingness is agnosticism, —an unmanly resignation and despair after a first defeat. The bravest, noblest attitude is that of unsatisfied longing, and the never-stilled faith that light will come into all of our darkness and that the riddle of our lives will be solved. We must trust that light exists, and must persist in trying to find it, must be willing to give up half-truths, for whole truths, and guard the mind against the bias of prejudice. Agnosticism, materialism, atheism, pessimism, sensualism, reckless luxury, despair, rigid creed-worship and superstition, and a thousand similar evils of our time, are directly begotten and fostered by this lack of trust in coming light.

The best method of reaching truth is that of a sympathetic study of the facts before us—that is, by induction, and by a broad, careful, systematic inference, by making the light about us ever wider, and fearlessly attacking the mystery beyond. Darkness and half-truths are thus progressively merged into light and larger truths, and the world of the known becomes ever greater and more satisfying. Let us renounce deduction, theories clutched out of the air, word-wars and logical juggling, and fix our minds to a perception of facts and a sympathetic interpretation of them.

Now when one looks about him the plainest, largest fact he sees is that of the distinction between living and lifeless things. All things are divisible into the two categories. Just as there is all the difference in the world to the mother's heart between the living and the dead child, so to the philosophic mind there is all the difference in the world between living matter and non-living matter. Upon that distinction must begin all honest thinking. A step further is gained by the obvious truth that the visible part of living forms is made up of the complexed, systematized, and utilized atoms and molecules of the non-living world. The invisible life is an organizing power using matter as material, and intelligence and purposiveness are plain characteristics of the invisible life. Every cell and organ and organism shows presently-acting intelligence, healing power, adaptation to circumstance, resistance to or use of circumstance. It is plain that a practically omnipresent, invisible, living, intelligent force is operating in and through every living thing. To identify matter and this living intelligence by any system of idealism, monism, pantheism, or materialism, is to do violence to logic and misread the facts. The dark riddle of life is to explain why Life is thus incorporating itself in material forms, and why the peculiarities, course, accidents, length of progress and evils, of the process are as they are. This little book, I believe, gives the keynote and method of solution of the riddle. The question of the origin of the lifeless universe is left untouched, with the acceptance of what seems to me the axiomatic truth that its essential elements, the atoms, are uncaused, did not originate at all, and, least of all, by any mental or intelligent fiat. It is a simple untruth that the non-living world of matter shows any faintest hint of design or of divinity. But there is comfort instead of despair in this background of fate, of a universe not made or ordered by mentality. Matter could not have been created, as something out of nothing is unthinkable; and only one law or mode of its action, gravity, is at present unexplained, and this "law" is really all the quality it has. The whole endless and eternal universe is made up of the same kinds of atoms as we know here. It could not have begun, or the despair-provoking certainty of its coming end would be inevitable, since, to use a simple illustration, a rope that has one end must have another. Nor can it be limited in extent, since any immaterial space beyond it, or any boundary to it, by the inevitable laws of gravity would bring about universal catastrophe. No such catastrophe is discoverable, or inferable from astronomic fact. Every large part of the sidereal universe is held more or less fixed in place by the opposing gravitational force of a literally infinite universe upon every side. If upon any side that universe were less than infinite in extent, the space-filling suns and worlds would be drawn toward a somewhere-existing gravitational center, and there would result a persistent and consecutive crashing together of all suns and matter into one tremendous central sun. Hence the attempt of astronomy to determine the permanent direction of motion of a body of stars toward a point, or their orderly revolution about a center, is useless and always will remain resultless. The existence of such motion or revolution would mean the speedy end of the extended universe. There is, therefore, comfort even in fate.

It is not absurd to ask as to the origin of life, because in the infinite progress possible to Spirit or Life, we may not block the way by any useless and impertinent agnosticism as regards even this great question. There are many reasons, deductive and inductive, for supposing Life to have had some sort of a primary inception, or progressiveness, however hidden from our present mental power. Every expression of Life we know shows process, difficulties unconquerable and difficulties conquerable, mastery by fate or ingenious partial conquering of fate—never a suggestion of omnipotence. The inference is quite clear, that if life were a worker in matter in all the past eternity, it would have been a more successful conqueror of it than is pathetically evident. The most patent aim of life is to win itself a home in worlds of inorganic matter, and to obtain progressive control of purely physical matter and forces. The fact that the success is only partial in our own world, that it has been attended with such difficulty and such expense (suffering, evil, death, reproduction, etc.), and that not more than two or three worlds of our solar system can possibly allow life a home in them, together with the certainty that like conditions exist everywhere else,—all this points to the finiteness, if one may so speak, of God, and His struggle with adverse circumstances. But it also gives blessed reasons and incentives for sympathy with Him, and makes duty clear, unravels a thousand mysteries of our being here, makes religion a psychical as well as a biological necessity,—indeed, forms the ground of an indissoluble and necessary identity of religion and biology. Lives must be linked with Life by love and sympathy and loyalty, just as much as they are derivatively, physically, and physiologically.

Upon its physical side this conjoining of religion and biology is made clear by the cell-doctrine which physiology and pathology have established as the fundamental truth of all biological processes. The physiological unit is the cell. This truth has been seen plainly enough, but only physically and physiologically; its application, as explanatory of the thousand mysteries of our own life, and especially as explanatory of what is called evolution, the process of life-incorporation or incarnation, and of the great mysteries of death, nutrition, psychology, etc., has not been seen. I have endeavored to suggest this application and explanation in these pages, and in more efficient hands than mine I believe the method capable of such splendid use and extension that through it men's minds may come to see purpose and hopefulness in our life, and look forward with confidence and trust to a future that before was blank with mystery and dark with death. According to this thought God can only reach incarnation through the cell, and the difficulties of cell-nutrition explain the necessity of what is called "evolution," the fact of development and the length of the process, as well as its thousand peculiarities, partial failures, and evils. The progress of the process to ever higher forms, humanization, and the spiritualization of humanity, are explained by the purpose of God, luminous with coming beneficence and resplendent with coming beauty; the delayed progress is due to the single fact that His only seat of power in the material world is the cell, of which He alone has direct control, through which all ulterior aims are realizable, and that the difficult nutrition of this cell-life is subject to a thousand conditions of temperature, food-supply, etc., and those accidents of untoward circumstance we call disease and evil. Accordingly death is the great unconquered difficulty of nutrition, and reproduction is an ingenious semi-overcoming of it. Accordingly the complex cell of the animal is only possible when fed upon the less complex cell of the vegetable. Upon the preservation of the conditions and perfection of the vegetable world is based the existence of the animal and human world. Hence the exuberance and pertinacity of vegetation. Civilization, with its railroads and scientific agriculture and multitudinous systematization of food-supply and distribution, for the first time in the world now assures a security and settlement of the nutritional problem, and permits the attention of God and humanity to be given to the work of spiritualization, to the beneficent and artistic use of power over the material world. A certain corollary of this great incarnation-process is that in veriest literalness, it is God's own life and self that lives in and uses every cell of every living thing, and that He is thus " very near " indeed to every one of us, if happily we but search and find Him. It is His omnipresent personality that constitutes the reacting psychical reality of our being; through the highest cerebral cells of sensation, receiving the synthetized, transmuted, classified facts or stimuli of the external world, and through the motor cerebral cells ordering the use of body and of the external world. Finally the certainty of our freedom comes from the certainty that we are the very sons of God, sharers of His personality, possessed of His freedom, and that He longingly awaits our educational fitness and loyalty to confer upon us the honor of inheritance of His kingdom, and of being His co-workers and friends.

A great philosopher said that two things especially inspired him with awe and reverence : the starlit heaven above, and the moral law within. Until I reached the vivid knowledge of the foregoing truths these two things were precisely they that inspired me with that utter desolation of despair I have called cosmic horror—that volcanic shuddering and sickening of the soul at the contemplation there without of the awful infinity of the dead, cold, and purposeless universe; whilst within, an unknown God, by an unknown instinct, commanded an unknown self to do an unknown duty. I have learned that many another sensitive despairing soul, in the face of the glib creeds and the loneliness of subjectivity, has also and often felt the same clutching spasm of cosmic horror, the very heart of life stifled and stilled with an infinite fear and sense of lostness. But I can now lie and look into the starry depths of space without soul-sickening or spirit-shudder, for knowledge lends comfort even to fate, and the certainty of the vision and the love of God in the world about and within me translates the stern command of duty into a sweet and irresistible invitation of the Father to help Him.

I believe that very few indeed find in the stars or in the command of duty either consolation or helpfulness for their spiritual homesickness and orphanage. As formerly with me, I judge that, with the common mental attitude, the two objects inspire terror and despair. There can be no doubt that from similar fear of the awful unresponsive infinity of the universe, and from the contradiction of internal command and external fact, unknown thousands consciously or unconsciously renounce thought and investigation, and either bury themselves in an authoritative system, creed, or church, or fly from the demands of duty to the distractions of money-getting, sensuality, fashion, and the thousand sad devices people have made in order to forget themselves. But the nobler can never stifle the soul's unrest and insatisfaction in such ways. The feeling can never be kept wholly out of consciousness that they are orphans who still hunger for a Father they have never beheld and cannot find, and yearn with a sickening nostalgia for a home they have never known and see no hope of ever knowing. More superficial minds turn to the hollowness of an absurd atheism and materialism or a more absurd spiritism, while the virile seek relief in science, literature, socialism, or government. But all alike have despaired and renounced, and have settled into a conscious or unconscious agnosticism that crushes the great decisive questions of life out of sight, and that logically and practically, sooner or later, leads to lowered ideals, more selfish practices and habits, to psychical, and not infrequently to physiological suicide.

The great need of the age is a scientific religion or a religious science that shall give these despairing, toppling minds a home in the universe; a religion that, blinking no fact of evil or truth of science,—nay, built beyond possibility of wreckage on the everlasting bases supplied by science, shall thereby have certain foundation for a superstructure wherein may joyfully dwell the Soul, with all its infinite possibilities of art, love, and faith. Such a suggestion of a reasonable and even necessary faith I firmly believe, however imperfectly, is herewith accurately given. For twenty or more years I have despairingly ransacked the wisdom of ethnic religions, systems of philosophy, and of natural theology, and lo! under the microscope I found God at work, and in biology revealing Himself so fast and so far as fate and His myriad difficulties allowed, self His incarnation and deputy, Duty and Intellect His pleading with the deputy to become co-partner and helper in the Divinization of the World.

The sole solutions of the modern tragedy of the spirit that so far are at all clear or worthy of consideration are those supplied by a religion that is not scientific, or by a science that is not religious. Science and religion equally and alike await the vision and knowledge of the ever-present, living, and struggling God.

The relation of philosophy and religion should and must be a genuine union and interfusion. They represent two methods of mental activity of one being, two aspects of one great life-truth. True personality will bind them into a living unity. The philosophy that is not religious and the religion that is not rational are either but half-truths, each needing the other to complete and perfect itself. The history of the marriage, yet, alas, to be solemnized, may be symbolized by the sweet story of Gilbert à Becket and his love. Captured during the crusades by a Syrian Emir, Gilbert and the daughter of the Emir fell in love with each other and planned to escape together. In this he succeeded, but she at first did not. At a later time she was able to elude her father and with great suffering travelled the long way to London knowing but two English words. For months, followed by a curious mob, she cried these two words, *Gilbert, London*, about the streets, until arriving accidentally before Gilbert's house, he by chance discovered her and took her to his home and heart. The more virile reason of the Occidental mind has escaped from the slavery of Oriental error, but Religion, that true and ever-faithful daughter of the East, has followed so soon and fast as possible, and now, seeking her lover, Philosophy, sadly wanders dazed and fainting about our modern world with but two words, *God, Love*, ever upon her lips, while the rabble follow with cold or superstitious curiosity, or cruel jeers.

It is the conjunction of scientific reason and religious emotion that alone can give the soul of man the chakar wings and gladsome spirit * to mount above the clouds

---

* "The heavenward flight of a large bird is always a magnificent spectacle; that of the chakar is peculiarly fascinating on account of the resounding notes it sings while soaring, and in which the bird seems to exult in its sub-

and storms of life when they become too oppressive, and soar high in the sunlit empyrean of freedom and happiness.

lime power and freedom. I was once very much surprised at the behavior of a couple of chakars during a thunderstorm. On a still, sultry day in summer I was standing watching masses of black cloud coming rapidly over the sky, while a hundred yards from me stood the two birds also apparently watching the approaching storm with interest. Presently the edge of the cloud touched the edge of the sun, and a twilight gloom fell on the earth. The very moment the sun disappeared the birds rose up and soon began singing their long resounding notes, though it was loudly thundering at the time, while vivid flashes of lightning lit the black cloud overhead at short intervals. I watched their flight and listened to their notes, till suddenly, as they made a wide sweep upwards, they disappeared in the cloud, and at the same moment their voices became muffled, and seemed to come from an immense distance. The cloud continued emitting sharp flashes of lightning, but the birds never reappeared, and after six or seven minutes once more their notes sounded clear and loud above the muttering thunder. I suppose they had passed through the cloud into the clear atmosphere above it."— HUDSON, *Naturalist in La Plata.*

CHAPTER I.

PHYSICAL AND METAPHYSICAL.

THE two essential constituents or factors of the work of every artisan—indeed, of every act or function of a living organism—are the material and the worker. This is the dualism that underlies all the outgoings of man's ingenuity and industry. There are many other subdivisions possible: for example, there is the finished product or manufactured article; but, on the one hand, that is evidently only the chosen and transformed material, or it is the idea of the workman incorporated in material shape; or, from the side of the workman, it may be said that his hands that as tools do the work, his eyes that aid him, his senses—nay, his entire body, are but instruments of his mind. His body is but the intermediate agent for the realization of ideas or ingenious energy. Thus the slightest reflection shows that all other phases of bionergy or life-force are resolvable into two fundamental ones: the material acted upon, and the intelligent energy that manipulates it.

Generalize this thought, and we have the necessary prerequisite of a proper conception of the world. All accurate knowledge, all valuable philosophic thinking, is based upon the same distinction of maker and material. In all the world there are finally but two objects, the physical and the extra-physical, or, in common language, Matter and Life. Perception, through the senses, with scientific imagination and inference, give us knowledge of the physical world ; self-consciousness gives us knowledge of the metaphysical world, supplemented by the observation of the results of Life's activity in the material realm.

THE PHYSICAL comprises all matter, the physical universe with all its inherent powers and properties, and forces, gravitational, and chemical. Herein, of course, is included the ether, with its forces, radiant energy, light, heat, magnetism, and electricity. The doctrine of the persistence and co-relation of physical forces, assented to by all rational minds, proves conclusively that this physical universe was without beginning, and will be without end ; that it is uncreated, existing in and of itself. The spectroscope shows that, in addition to its infinity of time, it has perhaps not a strictly logical, but what, without serious error, we may call a practical, infinity of space or extension. The farthest suns, the light-messages of which take incalculable light-years * to reach us, are shown to consist of essentially the same chemical elements as our own solar system. About these suns there doubtless are, as with us, swinging worlds hidden from our eyes by their distance and lack of self-luminosity, upon some of which in each system there can be little doubt there are millions of men wondering, as we wonder, about their fellow-men in other world-islands, swimming in the inpenetrable deeps of space. The doctrine of the atomicity of matter, of which there can be no rational denial (" centers of force,"

---

* A light-year is the distance travelled by light in a year of our time. It is equal to something like six hundred billion miles. Light travels about one hundred and eighty-nine thousand miles per second. A centauri is supposed to be the nearest star to our solar system. It is about four and a half light-years distant.

"vortex-rings," etc., are but aspects or modifications, not negations, of the theory), shows us that however combined into complex systems by mutual inter-actions or the master-forces of Life, they all tend to fall back again into simple systems or elementary independence whenever released from a mechanically controlling or an intelligent domination. Combine, dissolve, and recombine, however many million times you please, the atoms of the seventy-two or more elements that Life shapes and uses for purpose, and each preserves forever its independence, and is forever unchangeable, has been, and always will be so, wherever the universe extends.

THE METAPHYSICAL.—Geology and cosmology make it clear that so far as this world is concerned Life is a late-comer, and will depart early. Millions of years before the earth was cool enough to permit of a succession of living types, the physical held full sway, and millions of years hence, when too cold to permit life a habitation, it will swing on about the sun, a dead mass of cinder without the moulding and vivifying touch of its present wonder-working visitor. The inference that it is the same in other parts of the universe is beyond question. Thus it is conclusive that Life is a reality absolutely different from matter, and in no wise to be identified with it or derived from it. The proof of this fact is primarily an act of perception, secondarily of self-consciousness, but as we shall later see, it is also a mechanically deducible truth. The failure in perceiving this truth is the fundamental error of all monistic and solely deductive systems of philosophy, and has rendered self-contradictory and sterile the thinking of great and good minds. The fertility and light brought into philosophy by the sole sway of the reverse and likewise partial truth is shown by the results of physical science and Darwinism. However puerile and pitiable may seem the crass materialism of some of the sorry parasites of this school, one is astonished and delighted to recognize the beautiful banishment of mystery following the one-sided emphasis of the study of the material alone. The misfortune has been that by some strange fatuity, probably as a stupid psychological reaction to the God's-work-forgetting supernaturalism of a previous age, scientific research and philosophy too often deny all supernaturalism, and burden themselves with the farce of accounting mechanically for the mechanically unaccountable.

All of the usual terms heretofore in use to designate the extra-mundane source of purpose and intelligent power, connote attributes and sentiments that are either impossible to unite in such a being, or that are unjustifiably ascribed to him. Let us limit ourselves modestly to what is demonstrable or fairly inferable as to this being; by no means implying that our tentative boundaries are either final or definite. Let us stand always in the light and upon the known, enlarging so fast as we may our area of light and knowledge, but never jumping blindly into the black chasm of the unknown. If thus plunging we think to be upborne by the divine wings of faith we shall, as so many others have done, batter our heads with inglorious fatality against the rocks of ignorance. The God we see daily at work all over the globe is primarily and essentially *LIFE*. This word alone, for the ordinary uses of our writing and speech, is sufficiently full and rich to satisfy the sympathetic mind. But life to many means but little more than a mere blind energy, like Schopenhauer's *blinde Wille*, which, even to the concocter of that sad, brilliant philosophy, apparently had most fiery and piercing eyes. To the sympathetic and open vision, however, nothing is more patent than purpose, wisdom, and intelligence instinct in every living thing. To βιος, therefore, we must add λογος, and with the designation, *Biologos*, we connote all that is at present needed, while we leave unknown power and possibility unexpressed.

ATTRIBUTES OF BIOLOGOS.—The childish and uncritical faiths of mankind have in a fervor of wonder endowed their supreme Gods with a multiplicity of absurd, unjustifiable, and self-contradictory characteristics and powers, that have chilled true love and sympathy, and left the mind of the worshipper wandering in a maze of mystery and intellectual inconclusiveness, and his heart the home of sorrow and doubt. Much of this was because the so-called "natural theology" was oblivious of nature and its widest lessons, because men spurned or ignored the very works of God all about them, and of which their own bodies were the sublimest and divinest proofs. They plunged into an ecstasy of adoration and flattery, or of logic-chopping dialectic, that by no road led nowhither. The sorrowfulest and stupidest error of all was in the ascription of omnipotence to the being whose every plant, tree, animal, or man, conclusively shows struggle against difficulties, and ingenious half-victories, progressive failures and successes, devious and indirect results outworked with imperfect means and obstinate materials. This supposed attribute led thousands to doubt, denial, and atheism, by the simple and unanswerable logic of "Either He does not wish (to destroy evil) or He cannot,"—either horn of the dilemma, of course, negativing His existence. A part of the same fallacy was the like unjustifiable ascription of infinity to God,—an attribute wholly unthinkable when considered as a positive one, and leaving the poor befogged mind seized with sudden chill and amazed doubt.

As to omniscience, the thought is not destructive of religious feeling and intellectual clearness. We stand in rapt amazement at the intelligence displayed in the synchronous creation, re-creation, guiding, and governing, of million-fold types of billion-fold individual organisms all over the world, deep in its oceans, and high in its airs and mountains. Here is a science that so far as our seeing goes is almost an omniscience,—especially if we suppose Biologos covering millions of other previously dead worlds with million-fold forms of palpitant life. Perfect intellectual candor lets all doubts and qualifications have their full expression and influence, and therefore we may not quench the suspicion sometimes arising that, if Biologos were omniscient, He would have been able, even with far less than omnipotent power, to have spared Himself much indirection, much misdirection and waste, and His creatures the awful poignancy of wretchedness, which, like the groan of a monotonous sub-dominant, is heard in all the music of life. But there are a thousand illustrations daily happening in our hearts and before our eyes that all biologic phenomena are led and marshalled by an Intelligence whose ends, however hidden and misappreciated by us, are still very clear to Himself. We are much like soldiers whose primal duties are obedience, bravery, and confidence in the General-in-Chief. Even the highest of His subordinate officers does not know the entire plan of campaign, lying in the most hidden deeps of the Commander's mind. The last battle will make it clear. Moreover, the noblest of human minds and hearts are so fast reaching almost a practical omniscience, that it becomes easy to believe Biologos endowed with a knowledge of almost positive infiniteness. So approximate is it that the question has little significance, and its discussion is comparatively useless. The vital point is to recognize the sadly but awfully evident fact that knowledge with Him and with us has outrun power. We know much about the constitution and motions of the heavenly bodies, but are wholly powerless over them. Of the history and march of life-forms of our planet we have a growing accuracy of knowledge, but are able to modify or influence them scarcely any. In the same way the biography of every plant or animal shows that its wise Creator has been forced by indirection and inexhaustible ingenuity to obviate or overcome innumerable difficulties and dangers.

Precisely the same thought arises when we think of the goodness of God. To endow sensitive creatures with longings and possibilities scarcely less than infinite, and thereafter deny the bulk of them any proper realization of ideals; to make suffering and death the condition and means of progress;—all this, and more, seems incompatible with a Christ-like sympathy and a limitless goodness. But there are many proofs that we can see, whilst from them we may infer others not seen, which prove suffering and denied realization of longing to be either temporary accidents, necessary antecedents, or inobviable conditions of a larger ethics, in which the errant note of pain is blended as a passing discord to heighten the atoning beauty of the final resolving harmony. Slowly men come to learn that even now a large proportion of the world's suffering is obviable, and that each may attain a greater and at least a moderate happiness with a greater loyalty to God. Moreover, the purest vision sees that there is something better than happiness—happiness as usually conceived—always awaiting the purest heart.

Practical intellectual observation, therefore, harmonizes with the hunger of the religious heart in finding the Father-Heart of Life and the divine vision essentially the same as our own, transcendently larger and purer, willing the good, seeing and knowing almost or quite infinitely, but very far from omnipotence either physically or morally. Friends and parents, whom we love, are none the less prized because finite and faultful. Indeed we love them all the more if we see them ever struggling against difficulties within and without, and ever progressively actualizing their ideals. So is God brought closely correspondent to our yearning if we recognize Him as a finite and suffering God, of quite limitless benevolence and knowledge, but struggling with divine heroism against recalcitrant material and perpetual obstacle. Only such a conception can give a basis for religion, and make thinkable or possible, obedience to the command that we love God. No infinite and omnipotent being can be loved, as the Avatars, Incarnations, Virgin and Saint worship of all times practically prove.

At the first glance it would seem that this process of finitizing the divine were impious. To the truly religious it will not finally so appear. I am sorry for the dogmatist who will persist in his illogical logic, but for the professional dogmatist I have little concern. The time has passed when inquiry may be throttled by the authority of lordly domination either of systems of so-called philosophy or miscalled religion. And what is impiety? Primarily irreverence and unkindness. But it has often appeared that a loving trustfulness and a genuine reverence is sometimes found in the heart of the modest "heretic," that was conspicuous by its absence from the bulemic appetite of the "believer." Piety, I would believe, consists in the conscientious attuning of one's personality in the most perfect unison with the "tonic" struck by God for the world's orchestra. Intellectual honesty is a string often sadly out of tune, I fear. To the truly pious, I have said, the finite but the acting and loving God will be a welcome belief. Indeed, in all time the genuinely pious have made themselves such Gods when denied them by the articles of the dominant creed. Thus one God after another is created as logic and authority have pushed the previous God back into the soundless abyss of infinity. When Jehovah talked with Moses, and Himself led the children of Israel, no nearer anthropomorphic divinity was needed. But it was soon seen that the one-tribal God must perforce neglect the other nations. The God of all people, of all worlds, at once became dim and shadowy in His infinity, and out of the heart-needs of hungering humanity His Son is born to be near them. Him they can love. With blind fatuity the iron creed pushes Jesus into an equality of Godhood, robbing the human heart of its cherished faith. But the ever prolific, ever-hungering, again transfers its love to new semi-divine humanities, to the Virgin Mary, and to Saints innumerable. What of throbbing religious faith manages to keep itself alive in these days of practical sensualism and theoretical materialism clings with the energy of desperation to these pathetic fresh creations, at once idols and idyls of an undying faith periodically robbed of its young by the dogmatist and logician.

And likewise all over the world, new Avatars and incarnations without number attest the devotion of the people to a loving, suffering, acting, finite God, and the incomprehensibility and inutility of an infinite God.

The child's question, *Who made God?* is justifiable and pertinent. If the construction of our minds and the product of our experience makes us ask for a cause of things, it is just as necessary that we ask as to the cause of the cause. Alas, that the unwisest of peasants or the wisest of philosophers must render the same answer. In the light of modern science such a question as regards the origin of the purely physical universe seems highly absurd. In the whole realm of inorganic nature there is not the slightest hint of purposiveness or intellectual ordering. We may only ask concerning a designer when the product manifests design, but until Biologos created self-moving organisms with the dead matter of the physical world, no trace of extra-material influence is discoverable by the sane intellect however "God-possessed." It is unthinkable that matter could have been created *de novo*, and it is equally unthinkable that even divine power could annihilate a single atom. Matter and its properties, all purely mechanical and unchangeable, we must accept as eternal and self-existent. The sole considerable mystery in connection with it, of which we are still awaiting a solution, is gravitation, and possibly further study of the ether may give this to us.

But when we turn to a consideration of our metaphysical reality and that of Him who is the Father of us, we find no warrant of science or of immediate perception to justify us in a wholly restful certainty of conviction of His eternity or self-sufficiency. The perpetual continuance and infinite variety of ever renascent incarnations, every blade of grass and living thing being His Avatar, is a faultless proof of process and non-attained result. This alone is evidence of finiteness and incompleteness, a necessity put on Him from without (or worse still, springing from within) to be other, to attain an unreached end. Is there a hint in the restlessness and ever-uprising desire at the heart of us, and in the driving, almost frightful hunger of Biologos for world-wide incarnation, that there is also insatisfaction in the deep of His being, and that He Himself may also be a divine victim of some "struggle for existence." The thought may make us shudder as if icy-steel were in our soul, but every deep spirit has often felt the sudden sickening cosmic horror and chill as the infinite doubt of stability clutched his palsied heart while peering tremblingly over the crumbling precipice of supposed certainty into the abyss of past and future night.

Beyond all doubt, not matter-born, but matter-taming, Biologos came to our planet from without, whether, as has been taught, gaining the first foothold by means of a meteor-carried cluster of organic cells, or whether such an elementary organism were nursed and fanned into activity here in the warm ooze of some tropic shore, matters not. Life's organizing architectonic force is so profoundly unlike any mechanical force that the materialist of our day can only command our sincere pity for his congenital atrophy of perception. One who is blind to the extra-physical origin and hyper-mechanical nature of life-force is dead to argument. It is simply a question of perception, and to those who have never heard or seen, there is no mental image possible of sound or color. The fact of materialism, however, is a terrible proof of the power of prejudice over spontaneous and normal mental activity, whilst the pathos is heightened in that this materialistic prejudice is the reactionary child of a rabid monomaniac theology and mechanical creed-building.

By the most irresistible logic we recognize ourselves as incarnations or Avatars of Biologos; sharing His being we may therefore reason from our own nature to His. But since the subtlest and strongest human intellect can frame no faintest conception how intelligent life could have originated, we may feel some sorrowful certainty that such knowledge of Himself is perhaps hidden from Himself. But, on the other hand, man's ever renewed attack upon mystery, the torment of its weight to normal minds, is also a hint that even the child's intolerable question may finally get answer. The answer may not be at all what we should expect. Perhaps it may consist in recognizing an abnormalism of our mind begotten in us by our struggle with the laws of matter. To Biologos the necessity of ascribing cause to the ultimate elements of existence may not obtain. Have we an abnormal excess of the causation-ascribing impulse woven into our mental warp and woof by our million-year-long experience and work in the engendering of material forms? Uninfluenced by any of the material subjugations and changes, does mind feel any obligation to ask for final cause and origin? In view of the acceptance of the scientific doctrine of the conservation of energy and the indestructibleness of matter, we see how easy it has been for us to renounce the necessity of asking as to the ultimate cause of the physical world, and in this fact we catch suggestion that the mystery of the origin of the metaphysical may in time cease to have vexation, interest, or even significance for us.*

* A psychological inquiry as yet not thought of or attempted would be a pure effort to discern the validity and reach of intuitional or deductive truths as produced, modified, repressed, or distorted by the accidents of our total past and inherited experience in evolution-ages. But compensation for our insatisfaction in this colorless and airless height of dizzy thought is happily at hand when we turn to look at the silent, splendid cosmic panorama of biologic phenomena spread all over the world,—the glorious victory over death whereby the Father of Life has covered and filled the dead crust of a desert planet with soft, sweet, home-making vegetation, and with fleet-footed, limpid-eyed, and palpitant animal-life. Could rocks and fire, winds and waters, sunshine or star-shine, forces gravitational, mechanical, chemical, or ethereal, action or reaction, eon-long or world-wide, have resulted in a single grass-blade or cell of bioplasm? Fool indeed is he who hath said in his heart there is no God.

CHAPTER II.

PARTIAL TRUTHS.

THE truth and limits of materialism are simply the truth and limits of the physical universe, and rightly considered that truth is of such tremendous importance that the error and sin of Buddhism, of medieval asceticism, of idealistic, deductive, and pantheistic systems of philosophy, of most of the sentimental and theoretical religion of the day, are stupendous errors and sins. So ludicrous or awful according to the point of view are these errors, that the contempt of the scientist and of the eupeptic, if not justifiable, is at least comprehensible. And chiefly for the religious ascetic and religious idealist what consummate delusion and stupidity! These worshipped a divinity who, according to their own creed, made, ever upholds, and ever works in the world; and hence, in the fact of his own body, as in the magnificent array of cosmic animal and vegetable life, this impious religionist, if not prejudice-blind, must have beheld the dead-in-earnest work of his divinity and of his interest in this world. And yet he refuses to behold it, and ignores or hates the beloved handiwork of his divinity. Could folly be more foolish, or unreason more unreasonable? Could his deity be more outrageously insulted? What God considers worthy of interesting Himself in so deeply as he does in the laws and uses of matter, God's creatures may scarcely pretend to find of no interest whatever or to condemn outright.

Were his work but done in the spirit of loving sympathy, the physical scientist may by all odds be held to have chosen the wiser part. Alas, that too generally science is not pursued in such a spirit! If chemistry, physics, physiology, medicine, and forestry, could but catch the spirit of imitation of God, working as He works, and espousing His ends, what progress would follow, and what a light would come among their darkness! It is admitted that no life is possible in the sun, nor in the majority of his planets. On our globe but a fractional part of the matter forming its bulk has been or will be touched by Biologos. Moreover, for the greater part of its past the earth has had no living thing upon it, and will again refuse life a home. Biologos, therefore, makes use of but an infinitesimal part of the atoms and forces of the universe; the remainder pursue their own way independent and eternal. The delight and function of the intellect of man is knowledge, and, as an object of study, the physical universe is assuredly a large and sufficiently interesting part of all objects to admit neither of indifference to it nor of disdain of it. The delight and function of man's will is to do, and the sole material of artistic or useful recreation is the physical. The delight and function of man's emotional nature is sensation, and most of our feeling consists in the play of physical forces upon those exquisite wonders, the attuned and vibratile organs of the senses.

From a multitude of cogent reasons, therefore, there may be no indifference to physics and the physical universe. To find and maintain a safe home therein, and to use it wisely, requires all the assiduous study we can give to it and to its laws. With its clumsy but distinct justice it ruthlessly punishes ignorance or neglect of it. This constitutes the strength and justification of all materialistic or scientific modes of thinking.

But to hold that the material is all, to affect a belief in no other powers than normally inhere in mechanically ruled atoms and their natural correspondings,—this is a fatuous sin, exactly on a level with that of the ascetic.

PANTHEISM, "God-intoxication," has the merit of possessing a visual power that sees the most visible fact in the world, but it errs in supposably seeing God where He has never been. Not the millionth of the millionth of the millionth of the matter of the universe shows any trace of divinity. Moreover, pantheism finds God in the world by a process of ratiocination rather than of perception, and it is therefore a pale, bloodless God that it adores, a sort of ghost of a ghost. One has yet to hear of a confessed pantheist who also works earnestly as a biologist. The pantheistic divinity is so much of an abstraction, is so thin, so far away, and works in such subtle and indirect ways, that he is content to live millions of years as rock-strata, polar ice-caps, incandescent suns, or sterile planets. If the pantheist perceives God in the biological world he fails also to see what is equally visible and equally important, — His struggle with difficulties and obstinate material. But his chief sin is the topsy-turvyism of seeing divinity in the appallingly godless inorganic universe. In the universe of dead matter there are but two things: vibrating atoms and vibrating ether.* There are some seventy odd classes of tiny atoms, each member of a group exactly like his fellows throughout the infinite reaches of space, and each atom in constant vibration or revolution, according to rigid mechanical method and

---

* Perhaps more properly but one thing, as the ether is doubtless atomic, like an immensely rare hydrogen.

inherent necessity. The inter-revolution of two or more types of atoms in a closed system forms a compound substance, and the study of these constitutes chemic science. The rate of vibration, the extent of "free-path," and the relative crowding, constitute what we vulgarly call "heat," or solidity.

But in the densest bodies the substance is very porous; the atoms are always swinging. The thinnest substance of which we have any knowledge, the ether, passes through the earth much as wind through a leafless tree, and almost frictionless. Atomic immobility or absolute density, theoretically equivalent to—$273°$ C.,* is a purely imagined condition. These bunchings, clottings, or knottings of vibrant atoms constitute the suns and subordinate members of the world-systems. The strokes or beats of the vibrating atoms on the thinnest substance, the ether, give rise to resultant and similar vibrations and pulsings of this substance, whence the phenomena of heat, light, electricity, and magnetism. But always rigidly mechanical are these phenomena, always uniform and necessary, never betraying a trace of freedom or design. Language is a subjective convenience, and has no validity except that of mutual understanding and communication. If, now, we designate as mechanical all such phenomena of physics as these fixed, necessary, invariable, and inherent ones, language becomes meaningless and useless if we call them also theistic. This applies also to the error of monism. We simply meander in a dreamy waste of nowhereness and indistinction, if we smudge over the limiting uses of an agreed-upon nomenclature. Phenomena that show no freedom, purposiveness, reason, or rational use, cannot by any clear mind be classed

---

* The latest researches show that absolute zero recedes as it is neared, and that absolute atomic immobility is beyond this point.

in the same category as phenomena whose chief characteristics are spontaneity, design, adaptative reaction, and mentality. The mistake of materialism is to ignore the living worker; that of pantheism is to ignore the dead material.

MONISM is muddleism. It is the sole system of religion or philosophy without any truth whatever as a basis. Unlike any other system it does not exaggerate the partial truth and ignore the contradictory truth. Every system has done this, but the partial truth, of course, has given it validity and following. As monism has not done this it therefore has little or no validity and following. It contradicts the fundamental law and function of the mind, a perception of difference in two things. It contradicts the first result of action of this fundamental mental power, in that it asserts that life and matter are essentially identical. In which case, to be sure, language and the naming of things become useless, and the only logical thing for a monist worshipper to do would be to sit staring at his navel, and for all eternity monotonously murmur, *Om. Om.*

It is perhaps possible as a mere feat of logical or intellectual gymnastics, not unmixed with a strong touch of legerdemain, to construct some colorless concepts that cover and seem to unite matter and life in a pseudo-identity. But after this is done nothing is gained. It is quite as reasonable to start out with two unknowns, or infinites, as with one. Since time began, the weary brains and fancies of thinkers have tried to solve the question of ultimate origins. But all are agreed that no solution has been reached, and that much good brain-power has been wasted in the futile effort. Let us then leave such resultless searchings, and accept the two indubitable facts that stare us in the eyes from every object of the actual world.

If life and matter are by and by to be fused into an harmonious unity, the truth, when it comes, will be plain enough without our frightful labors. I do not see why a tentative dualism should be considered plebeian and infidel. On the contrary, a frank and free vision sees as the most certain of all facts a purposive and intelligent power, life, using and dominating a wholly alien, unsympathetic, and changeless material, physical substance, and no mental ingenuity can imagine either as derived from the other, or even as having any origin in time.

It is wholly useless to struggle against this dualism with an inherited bias of mind for a divinity responsible for the physical universe. Any system of monism, dualism, atheism, materialism, or idealism, can by an astute and erudite dialectician be proved or disproved with incontrovertible logic, provided that that logic be deductive in character, and that certain axiomatic premises be admitted. I have little or no respect for such logic, but ask only as to the fact that exists here before us. In the physical universe untouched by life's transforming power, there is not a trace of evidence that any but mechanic forces have ever moved an atom. But in all life-dominated matter there is—even in every atom's movement—something hypermechanical. If the responsibility of God be extended so as to comprise the inorganic universe either as to origin or as to present dominion, human freedom is then a myth, determinism is unavoidable, religion is nonsense, and the existence of evil inexplicable. In biology there is nothing more certain than the existence of process, unattained result, purpose being wrought out. Now process, *ipse facto*, demands a non-omnipotent God, and it confuses beyond all sanity to extend the reign of a finite divinity into the origin and government of the entire inorganic universe.

This is simply to locate fate, or the limiting and controlling power of God, behind both Himself and the inorganic universe. Process demands a finite God, and the fact evident in every biological phenomenon is the partial, difficult use of dead atoms and mechanic forces by an intelligent life-force.

Nothing is gained by seeking to go behind the patent fact, the certain dualism of life (or God) and matter, united only in the biologic process. Seek to unite these two forms of existence anterior to the biologic process, and intellect is anesthetized, and the mind goes mooning in a haze of dream out of which can come no clear decision either of reason or of morals. As a matter of fact, every great religion or system of ethics has been rooted in dualism. No human action can be ethical that does not consciously or unconsciously spring from it. If one denies the fact of process and progress in biological history, he may logically enough soar into dreamland and wordland, and rest content with the imaginative phantasmagoria that Berkeley, Hegel, Spinoza, and others have spun there. It seems to me an immoral desire on the part of many minds to ignore the evidences of struggle, process, and endeavor, most awfully present in the living world, and to find God in the mechanic universe as omnipotent author and governor. It is both intellectual sin and moral sin, because at one blow it palsies virile thought and kills sympathetic endeavor.

Lastly, it is the most awful impiety; because, unless one absolutely deny imperfection and unattained result in biological history, any monism puts the reason and ground of this evil or imperfection in the very being of God, instead of in a condition outside of Himself, and which He is successfully and progressively overcoming. Waste worlds, incomprehensible mechanicalism, glaring evil, deadness, uselessness, are everywhere the distinguishing characteristics of the mechanical universe. If God did all this, He is an incomprehensible God, and covers in Himself such hideous contradictions that I prefer blank atheism as being far more truthful and more comforting both to the intellect and to the heart.

And such an attempt to hold the human mind to an acknowledgment of the divine origin and control of the inorganic universe has turned thousands and millions to atheism. The intellect denies such a God the warrant of fact or of logic, and the heart rejects him as a kind of jumble of Fate, Devil, and God, the compassionless deity of all pessimisms and self-deceptions, and the fatalistic servant of his own imperfections and contradictions. Distinguish! Fate and Evil and delayed progress exist, their great foundation being the uncaused, non-intelligent, inorganic universe; Intellect, Freedom, Goodness, Life, and Love exist,—all conjoined in the God we see working under difficulties, and with the inorganic universe as dead material. Conjoin these two types of fact or existence prior to biological history, and no mind thoroughly comprehending the absoluteness of the logical contradiction could act or persist under such a condition, and the heart that loves goodness or beauty is at once chilled to icy death. If one prefer to make a finite God, a non-omnipotent struggling divinity, the cause and controller of the whole inorganic universe, I see no advantage therein over the dualism of life and matter that I urge, and that confronts us in every biologic fact. With the latter belief the mind is free, the heart religious, and sympathy springs unbidden for Him who in and all about us is heroically at work against intractable material and tremendous difficulty. With the former belief the facts are at variance, and from a tragically perplexed mind and heart, doubt and despair spring unbidden.

BUDDHISM is the most pathetic fact in the world's history. It is the heart-broken confession, first of one of the largest minds and hearts that has ever lived on this earth, that God's tremendous labor of incarnation and world-revivication is a failure and must be renounced. In the acceptance of this conclusion by so large a proportion of the human race, the confession gains a tragic grandeur and awfulness that reveal the terrible struggle of Biologos to perpetuate the human race, and for centuries how terrifyingly near catastrophe He was. As Buddha and every Buddhist was nothing less than incarnate Biologos Himself, in specialized and functional form, the solemn tragedy reveals the very face of the suffering, half-failing God.

But while this glancing aspect shows the long suffering and plaintive side of failure and abnegation, we must not forget that it is only partial, and that it is the average, the massed expression of the "composite photograph," that reveals the permanent and essential expression of the divine physiognomy. If the great world-drama was a failure in India and in Buddhism, elsewhere the hero's face shone with virility and victory. Greece and Israel were ripening while India decayed. Here was youth, while there, the *machinery* of life seemingly obtained the control and exhausted the ingenuity of the engineer. Blind forces got the upper-hand of spontaneity, habit crystallized into dead routine which the divine freedom and vitality could not break up. Life got imprisoned in formalities and class-interests, the sense of play and victory faded out, and for that people and time all that the keenest intelligence could see was eternal pain and failure; all that the most heroic ingenuity could devise was a way of release by return to Nirvana's renunciation of desire and incarnation; all that the sublimest love could do was to sacrifice power, pleasure, life, ay, even Nirvana itself, until all others should first have entered its blissful portals.

A powerful factor contributing to this pessimism was the unquestioning acceptance from Brahminism of the doctrine of transmigration. It seems strange that this race of consummate metaphysicians should have developed this doctrine to such extreme exaggeration, and that it should have become so woven into the very texture of its psychological structure. The reason of this, of course, is that it is such a profound truth. No great belief of a large body of people can be without a basis of fact. It will always be found to rest upon a partial truth. The greater part of the doctrine of transmigration is the truth that under a different name we have locked up in the mystery of heredity,—a continental mystery upon the shores of which some of the discoverers of science have begun to try to land. The incomprehensible phase of the Indian belief is the haphazard way the disembodied spirit gets its new incarnation. In a universe to them most orderly and rigidly regulated, this new housing of the freed soul appeared, as it seems, to be pure accident. One must not kill a flea, because it might be the soul of one's grandfather.

This element of chance in a world so law-bound, and in so weighty a matter, gives us pause and excuse for reflection. It certainly has this much of significance, that it shows recognition of the element of chance and accident, recognition too by deep minds, keenly metaphysical, and by millions of them, through thousands of years of history. To the pompous scientific Louis Fourteenths of our modern type of thinkers who are ever harping on "the orderly government of all phenomena by unbreakable law," and such like bosh, the conviction of the Oriental mind may have no disquieting effect. It is a thankless task to say to them, first, that "law" is never an *ab extra* superposition of method upon facts, but is merely the statement of the inherent method of process of phenomena; second, that even in physics the fact of one happening falling upon the fact of an unrelated process produces a chance or accidental sequence. That a fern leaf in a coal-seam should lie east or west is accident, and that the coal stratum should "dip" in one direction of the compass is due to unpredictable or accidental causes. Moreover, in biologic phenomena the mechanic rigidity or lawfulness of inert matter seems to impress upon them almost all the legal method they take on. The only laws to which there is no exception as important as the law itself, is some large generalization or colorless abstraction that leaves spontaneity and accident an adequate play-room. In the realm of free spirit, as in the pre-incarnating Biologos, and in artistic creation, in the sweep of dream and in the flashing of imagination, freedom knows no law and laughs at prevision.

The sin of intellect of Buddha and of his followers consisted in non-recognition of the work of Biologos, and in not perceiving the living God behind and in living things. They had no penetrating power to see this vision of the unseen. The distinction between the living artist and the dead material was not clear. The sin of character or emotion followed naturally on this intellectual sin: they were disloyal. Biologos never gives up; He is never indeed altogether conquered; He is among Polar ice mountains, in ocean depths, and even in the trackless deserts created by man's improvidence. He is always trying to win the world to Himself. The first of all human duties, the root-principle of all duties, is loyalty to God. Had Buddha but studied God's way and Being more accurately, had he but been His loyal acceptor and helper, the unexampled power of his influence might have won India from caste and despair, and rescued her from a history of submission and alien domination. But to Buddha there was no God, and therefore no way of God. Cycles of formalistic and dead religion had quenched in the hearts of the people the perception of the supernatural; hard habit had hidden from them the conception that spontaneous intelligence ruled even biological phenomena, *as much as it could,*—and, with the fading of the perception of God, came in the Godless religion,—although, to speak more properly, Buddhism is but a people's philosophy of method of gaining release from the thraldom of incarnation or material form. Seeing no Father behind the veiling of organic events, it naturally came about that the little group of fatherless children must bind themselves together more earnestly in brotherly love and mutual helpfulness ; wherefore arose a missionary spirit, a purity and a tenderness of heart, and a self-sacrifice, rivalled nowhere unless by the early Christians. The fact is a magnificent contradiction of the shallow nonsense of some dogmatists who have stoutly averred that a lively degree of morality and love cannot exist without a theistic belief. If Buddhism may be called a religion it is certainly an atheistic religion.

Born out of Brahminism, the dying sigh of a failed civilization, Buddhism, if not justifiable, is at least explainable, and commands our sympathy and pity. Not so modern pessimism. In the Europe and America of to-day fruit awaits the picking, opportunity comes to meet virility, and positive badness in a child is less repulsive than whine and sulk. While the hibernating Schopenhauer was grinding out his ursine philosophy, the brave Goethe was singing imperishable songs, his eyes agleam with the light of the rising sun of Science. If Goethe's craniology and his optics were as errorful as Schopenhauer's phantasmagoric idealism and diabolic will, the faithful one was right, the faithless one all wrong. Had the disloyal one but devoted his splendid power of psychological analysis, the misused verve and brilliancy of his incomparable intellect, to the service either of God or of man, what a helper he might have been! Instead of that, a hinderer, disloyal beyond compare. And one sin always treads upon another's heels:—From the strong, headstrong thinker have sprung a sorry brood of croakers and maudlin sentimentalists, whose puerile pessimism is little else than the voice of *ennui* and exhausted sensualism. These echoes of the feigned despair of a perverse but virile thinker are without every attribute that wins for their master some measure of reluctant admiration. They are belated births, misplaced in a world where success awaits the thinker and worker of every type.

CHRISTIANITY, it cannot be denied, has failed to conquer not only the world, but even its own communicants. In great part this arises from the indefiniteness of its basic creed and of its essential characteristic. There is neither logical unity nor common assent on the part of those calling themselves Christians as to what the ideals and aims of Christians should be. An unbiassed ethnic student can find little or no warrant in the words of Christ for the course of Christian history, even when the church was most powerful. The motive force of St. Paul's fervor was beyond question the belief in Christ's resurrection from death and his literal second coming. The manufacture of a trinitarian creed and the establishment of a Roman Catholic church with its pomp and ceremony were as little dreamed of by the Galilean carpenter as the truths of spectrum analysis or as the nature of civilization in the planet Mars. If Christ's teaching means anything the whole history of medieval asceticism was unchristian, and for a thousand years the bitter warfare of the Inquisition and of official Christianity against the irresistible on-coming victory of reason, freedom, and science is, if proof were needed, the most powerful proof, either that Christ had not christianized his Christianity or that the human mind must seek its religious incentives, ideas, and control outside of Christian precept and example. The division of the church into Roman, Greek, and Protestant, with the ludicrous schismatization of the latter group, each set at heart thoroughly convinced that it alone represents the founder, leaves but two alternative conclusions to the dispassionate student: either this "Gospel" of Jesus was too indefinite to cohere into a body of authoritative and understood truth; or his personal influence was incapable of impressing upon his followers this truth with sufficient force or momentum to make it carry beyond the enthusiasm of his little band of personal followers. The undebatable deduction is again alternative ; either that by thousands of years his teaching was superhuman, in advance of his age, beyond the comprehension and actualization of his professed followers during all this time, or that it was a tremendous error. From every logical *impasse* we are forced back to conclude that individual judgment, based upon entire lack of prejudice, upon faultless erudition, and after at least many years of the deepest investigation and study of original documents of the entire history of all branches of the church, and especially of the reported works and life of Jesus,—that individual judgment must decide what to do about the acceptance or the rejection, as a whole or in part, of the Christian church, or of the "Christian revelation." Nothing less than this will satisfy absolute intellectual honesty. This only a few scholars can do, and it is sad that the few that have done it reach non-unitary conclusions. In the hurried sweep of our little conscious being through the cycle of life's changes, interests, and duties, the majority can only look hastily about them, a little way before and after, catch here and there a maxim of wisdom, a cheering word, the glimpse of a precious example to follow, the hurt of a sin to avoid. Life is certainly for action, emotion, and perception, not for exegesis and biblical criticism. Christ flung his undying message upon the winds and waters of world-accident, and all may find it if they look for it rightly and carefully, bewaring of the organizations and creeds that would officially minister it.

I opine that the God of Christ was far wiser than the spirit of Christianity. Jesus, by a wisdom that seems superhuman, or by a fervor of emotion that gave him no time to organize, left no written word or authoritative statement. The circumstances of his life also conspired to the same end. The pathos of the world-wide gulf between himself and his followers becomes awfully tragic: " Could ye not watch with me one hour?" Through the dark night of those surroundings, the impenetrableness both of crucifier and follower to such ideas as his, there flashes the repetitive lightning of a divine wisdom and a divine love that strove equally and vainly against the sin and the formalism of his followers. In the glow of his death the reflection of these silent lightnings could certainly never have reached our eyes had it not been for the accident or for the miracle of St. Paul the persecutor—one who never saw Jesus. It is indeed most weird and wonderful to note how, again and always, the significance and preciousness of Christ's message was passed on by hands and hearts that knew it not. In conversion and life-long missionary work Paul was dazzled by it rather than understood it; creed-makers juggled with it; ecclesiasticism stole it and built upon it; sects wrangled over it—all fascinated by it, but all unconscious of its vital import and content. Such are the ironies of history.

But more mordant still becomes the irony when it is seen that this unappreciated message of Christ has been carried by those it was destined to destroy. It was all the while at work upon and against the hearts and institutions of its defenders and carriers. The most unchristian and antichristian facts of the world, the Nicene creed, the Catholic church, Asceticism, the Inquisition, Versailles, Bourbonism, and Calvinism,—all have prided themselves as being defenders of a faith that in very truth was secretly opposed to them and that was slowly, silently, undermining them. Had they but realized its significance, how they would have hated it! Thus God compels messengers to convey to new masters the sealed orders of their own execution.

But pride not yourself Churchman, Catholic or Protestant, Orthodox or Heterodox. You are also ignorant, sinful, unconscious bearers of a message you understand little better than your official forefathers. You also carry the order to a better Master of the Future for your own death-sentence. You, in your way are quite as ludicrous an exponent of the gospel of father-love and brother-love as battling Hun-Bishops, Torquemada, or Pompadourist. You have not seen the silent, searching, tender eyes of the yearning invisible Christ, waiting through centuries until by your ignorant hands his word shall at last reach the receptive sympathetic heart that knows, that understands, and that executes. You are still not his knights, but servants still of the sin of all centuries, the sin of luxury and of self-comfort. Nor is the illusion ever quite complete in your own consciousness. You cannot quite deceive yourselves or others. There stand the words, gloss them as you please, and ignore them as you may. There also are the men your brothers, they who vicariously minister to your physical comfort and who vicariously suffer for your sins. There also are the untaught, mistaught, sin-taught children, there the vicious and the filthy, there the corrupt politician making your laws, there million-fold parasitism, there the habit-enslaved, the insane, the poor, the diseased. From every page of the Jesus-records there blazes and burns the unquenchable fire of social love, and the duty of socialism: " Go sell what thou hast and follow me "; " Inasmuch as ye have done it unto the least." Whilever there exists injustice, a wronged, suffering, sinful, or ignorant one, it is only the sham Christian that can enjoy ease and luxury, and that can build or have aught to do with luxurious churches. Luxurious, time-wasting, and time-saving " worship of God " is in Christ's eyes the veriest acme of hypocritic sin. Foolish folk compound with their consciences by building useless churches, instead of inquiring why the more churches there are the more empty they are.

RELIGION, the mere fact of its existence, and especially of its tremendous rôle in and before all history, if anything could do so, should give the lie to materialism and determinism. It confutes the one and confounds the other. Here is a somewhat as evidently *ab extra* as it is lawless; and yet it is everywhere found dominating new nations and enslaving old ones, the cause of innumerable wars with weapons or with words only, an agent that turns all forecast topsy-turvy, an ever powerful fact at least, for good or for evil. There is much latter-day increase of the unreligious, and of the irreligious; of the agnostic, who under a name thus euphuistically seeks to conceal his atheism and materialism; and of his Philistine-brother, the so-called "scientist" who thinks "science" conterminous with physics, and whose brief creed is, No God but Evolution, and Spencer his prophet.

These logicians most stoutly maintain that all things take place according to invariable laws, the chief of which is the law of cause and effect. If so, then whence and wherefore religion? But when it does not suit his purpose, *i. e.*, the support of his prejudice, there is nothing this excellent logician cares for less than for genuine logic. Belief in an extra-mundane cause, a belief dominant in all the affairs of men, from savagery to civilization would seem to mortal and human logicians to imply the existence of that cause itself. If the predominant factor of all evolution is the interaction of organism and environment, "the reaction of sensitive protoplasm to external stimulus and change," as we are tirelessly told, then whence and where the "external religious stimulus" to which the organism has been so frightfully and persistently reagent? The farcical attempts to explain the religious belief of the world as ghost-worship and ancestor-worship are pleasant reading for the psychological humorist. It goes without saying that in a world ruled purely by inexorable law the simplest religious idea could never have gained admission.

But if religion and duty have been the twin voices of God, why is it that they have so often spoken the commands of the devil? The fact is most undeniable. The answer is threefold: 1. The rôle of religious ideas in the world has in great part been that of psychological tutelage, a metaphysical gymnastic exercise for the mind, against the time when it should undertake the real drama of rational, religious, and ethical life. All processes must have initial and intermediate stages, and the world process is no exception to the rule. 2. The respect of Biologos for the freedom of rational creatures and their development toward a freedom like His own, presupposes a schooling and apprenticeship in virtue, the perpetual power of choice, the tentative trial of evil in its thousand-fold protean disguises. Any government of His that would interfere with the freedom of man would defeat the progressive cultivation of spontaneous ethical and rational activity, itself one of the ultimate ends of the biological world-process. 3. However crude and apparently far from the highest religion, the general religious status of a tribe or nation is that for which it is adapted, and which corresponds to the phase of its development and of its modifying and imperative necessities.

Any religious idea or custom whatever stands as incontrovertible proof that biological phenomena are not solely those of mechanical action and reaction, that sensation and nutrition are not the only laws of organic development. I conceive the origin and development of religion in human history to be that of a free discovery and an inference, rather than that of an imposed command; religion had its origin within rather than without; it is a discovery rather than a command; a progressive educative finding of the Father, rather than any revelation from the Father. All truth is acquirement and reward of searching, religious truth certainly not excepted. Christ's revelation, his intuition or finding of religious truth, the best and purest the world has ever listened to, was, from present standpoints, sadly deficient, and in some respects it was erroneous. Many things that sound strange to us may be due to the misreporting of his cloudy-minded, unappreciative followers. That indeed is a sad possibility in reference to every line of the gospels. A hundred sayings show the great mind adapting the great truth to the almost infantile comprehension of his childish audience. We must, of course, throw out the whole disturbing illusion of myth and miracle that fond love always weaves about the persons of its adored religious teachers. The doctrine of non-resistance; the teaching of his literal second coming to judge the quick and the dead; the non-interest in government, good or bad, implicate in "Render unto Cæsar, etc.,"— these and many more are instances of positive error. Of things omitted, it may be noted that no word exists from him as to kindness to animals, and especially as to the tremendous importance and truth of God's revelation of Himself in vegetable and animal life. Thrift, a primal duty of man, is in many precepts and instances distinctly scorned or discouraged, whilst of family duties, sexual, and other matters, there was a negation or an aloofness that would but little help poor laboring men and women in their trials and temptations. He strove, of course, as is plainly seen, to give his followers a new birth, to make all their living spring from a new motive, expressed in the eleventh commandment, the essence and summarization of all his teaching. Thus acting, all minor rules would be unnecessary, and all questions of subordinate ethics would at once attain solution.

Being a Jew there could be no doubt and no merit about the theism of Jesus. There is personality and divinity behind material things—the truth that Brahmin, Buddhist, and Confucian failed to see. The unique teaching-service of Jesus was in the sole command that we love instead of fear Him, supplemented by that other * that all men are to be loved as one loves self. And certainly no human wisdom has ever reached a more compact, comprehensive, and enduring summarization of religious and ethical truth.

There is nothing more remarkable in all history than the persistent and almost universal practice of the human mind of religion, and the belief in the existence of a hyperphysical governing intelligence. Against the obstinacy and unresponse of the inorganic world; in the face of death, and what was worse, of life-long suffering of the innocent; in the face of vicarious pain, of unmerited and inexplainable poverty or luxury, man still believed in a ruler and a cause of living things. He did not care a fig for the illogicality of the belief with which the atheist quite logically mocked him.

What was the function of this belief? It corresponded only partially to the truth, and it gave him infinite suffering. Why should such a true-untrue and inexplicable thing have dominated his " environment-reacting " organism? On his part it was a perception of fact that his intellect had not the strength and training to correct and limit according to the teachings of things, and yet it was a belief too precious and true to renounce altogether. May we not say that the function of the belief was to preserve the

* Both quoted and acknowledged as from Moses.

essentials of the truth, to keep the tradition until such time as living perception and vision of the whole fact should become possible? It was the assertion that God existed, however much determinism and godlessness seemed to prevail. The process from pure animalism to a stage of civilization hardly yet reached, was one that would every day have ended in disaster without the saving fact of hope, and the illogical conviction of final victory. Viewed simply in the light of logic, the readings of the day's signs and the deductions of the day's facts could end only in an acknowledgment of failure and in a renunciation of endeavor. Buddha drew the conclusion with infallible accuracy. But they reckon ill who leave God out of the count, and God *is* in the count despite all the Buddhas and atheists of all time. Religion and its everlasting twin-deity, Duty, held men's faces to the east until such time as the sun should rise. It still holds them toward the sun until the surgeon of science shall remove the cataracts of ignorance, of habit, and of prejudice, and they shall see God. Dazzled and dazed the scientific mind is at present like the aphakic, suddenly brought to see, but not recognizing or knowing what he sees. It still sees men as trees walking, and does not know that what it sees is at last the benignant and beckoning God himself.

Religious belief was thus and literally a "saving" faith. It kept men from despair. It was one of the safeguards the human mind created to carry it through the process of humanization. Without its staying and supporting, the monkey would never have become man, nor the man philosopher. It was the *sine qua non* of civilization, and to future species, it will certainly have *this* much of validity.

LIMITATIONS AND SUPPLEMENTATIONS OF THE ELEVENTH COMMANDMANT are necessary, even by those accepting it most heartily. In almost any statement of any truth, it is quite possible, and upon the philosophic thinker quite obligatory, to find a large measure of error or of exaggeration, and, in the absolute reverse of the statement, a large measure of truth. This is the sufficient basis of failure of all monisms, deductive systems of thought, of all one-sided idealisms and one-sided materialisms. Hence the stringent necessity of testing every statement by experience, or by the first-hand study of facts. The human mind has been produced or developed as a reflector, a perceiver, and a moulder of facts, and therein lies its beauty and power. Just here arises the criticism of almost all religions; they are deductive systems: they superimpose theory, statement, or emotion, upon the facts; they are not the products of experience, but they are judgments or ideals antedating experience. Hence they are often found to need correction or outfilling by experience.

The law of love of God and man, as true and as binding as any law can be, still needs interpretation by the living intelligence, needs adaptation to the needs and ideals of the individual life, and of all living beings collectively considered. In such interpretation and adaptation it will be found that its absolute sweep of application will suffer limitation, and that always and everywhere it is not sufficiently comprehensive to furnish a rule of life. Such a primary limitation is the truth that the connotation if not the denotation of the word God, according to all old habits of thought, is omnipotence or absolute creation and control of all phenomena, organic or inorganic. But it cannot be represented in thought, and it never was observed in fact, that the physical universe was either created, or in any least way (except through the great incarnation scheme) controllable by God. To love an omnipotent God is psychologically as impossible as it is to represent such a being in thought. No faintest thrill of tender emotion ever went from a human soul to a God clearly conceived as omnipotent. Whatever the creed and the theologian demanded, the wiser human heart, blindly knowing a better way, always anthropomorphized his God, and this process consisted essentially in leaving out of sight and out of thought His supposed omnipotence. To have the power and not to help, chills every possibility of love in a normally made being, and when, as in some aspects of Mohammedanism, Judaism, and especially in Calvinism, the thought of omnipotence gets comparative domination, God at once becomes a sort of devilish determinism, or grinding machine, whom one may fear, but from His worshippers even His omnipotence could wrench no spark of love. Christ should therefore have given his disciples some hint that God does wish but is not able, rather than that He is able but does not wish. There is no Fatherliness in the "Great First Cause, least understood."

Religion, in common with all other activities of the human mind, has suffered from man's general indifference to the fact and significance of the world of living nature that surrounded him and of which he was outcome and part. Standing before the wonderful panorama of nature, man, in the first place, confused the living and non-living into one indiscriminate jumble, and thus prevented himself from seeing the evidences of intelligence, design, and tendency so clear to our eyes in the world of living forms. In the second place, he had not yet freed himself sufficiently from the egotism resulting from the "struggle for existence," to be able to perceive an objective fact except in relation to his own selfishness. As a consequence of the necessity he was under of ruthlessly dominating and using everything that could be of service to him in order to secure his own individuality in the thousand-fold warfare of his daily life, things that had no relation to that service became as if non-existent, and only just that aspect of the serviceable thing that yielded the service was known or attended to. Hence until now that which was neither tool nor food nor enemy, was ignored. Even now that insane fury of clutching more than we can either eat or use, called money-making, and the shooting of animals for "sport," are survivals of the barbaric struggle, and show how little we have outlived the cruel time. Hence the essence and scope of religion was the relation of the human personality to an extra-mundane personality. The study, knowledge, and love of God, as manifested in organic life, was practically unknown.

The points toward which I have been looking in a too long introduction are that not only can one not love inactive omnipotence, but one cannot love an inactive God of any kind. There is hardly any love of God psychologically possible except as one loves the work that God is doing in the world. There are numberless reasons for this fact, among which may be mentioned the evident one that solely in the affairs of men, one's fellows in time or in history, there is no striking evidence of God's existence or of His government,—such evidence, I mean, as the unerudite and unphilosophical could find or appreciate. Moreover, without nature-love and knowledge (by which of course is meant of the living phenomena of nature) God remains a sort of *ab extra* cause, an impersonal somewhat, or an unknown power. There is no adequate or hardly any partial realization of the Fatherhood of God except with the emotional recognition of the fact that he is not only *our* Father, but, in as exact and literal and loving a sense, that he is also the Father of every grass-blade, plant, fish, insect, or animal throughout the whole wide world. Through insight into the intent, character, and work of Biologos, one may come to some appreciable degree, to know and to love Him. This will form the basis for the oft-talked-of religion of science, or of the future, but no clear eye can look into self or into his fellow-Christians, and fail to see that love, that spontaneous play of emotional trust and cherishing, is not possible toward an unknown, impersonal, and vague " Great First Cause," such as too often God has been represented. No wonder that the study of "secondary causes" and "influences of the environment" seem to cheat God out of his slender share in man's life, and that to the " advanced modern thinker " the world seems godless. So unreal a divinity is easily displaced from a throne so fragile. Men respect the phrase in lesson and reading, they hear and assent, but they do not love an unknown God. It may be said that the soul knows its Father directly, that personality touches personality instantaneously, and, without either a knowledge of history, of men, or organic nature to intermediate or to aid, it adores and loves Him. I do not deny it, and I am glad to believe it, but it must be confessed that it is a highly exceptional fact, and that, far from what one would naturally expect in such cases, the direct appreciation and immediateness of union leads to little or no secondary study and love of God's work in nature, whilst the road " through nature to nature's God" is far more beautiful and more often travelled.

To these and to many other reasons that might be mentioned, may be added the fact that the placid love and study of objective nature is necessary to win the mind from the egotistic and selfish habit of all previous time, and to permit the recognition of the large impersonalities and final ends of biological history, toward which Biologos is calmly working with individuals, types, and civilizations, as his tools and his stepping-stones. The humility of true science outdoes the humility of religion.

To the command, Thou shalt love God with every fiber of thy being, there must therefore be added, in order to make the command possible and effective, Thou must seek to know God, sympathizing with Him in His aims and in His difficulties as shown in a discriminating love of His work in living nature.

As to the second part of the Eleventh Commandment, it is a perfect ideal of human conduct, but like all ideals, of course, almost never realized, or realizable.

The whole command, indeed, partakes of the unattainable that characterizes Christ's life and the lives of all those who have approached Christian perfection, or who have been carried off their feet by the unearthly enthusiasm of Christian feeling. A sort of passionate somnambulism seems to be the characteristic of strong Christian feeling. Jesus seems to have walked in a kind of dreamy rapture, oblivious of many homely but distinctly necessary virtues, scorning thrift, frugality, legislation, government, schools, settled habits, and the work-a-day world. One wonders just why he permitted himself to be crucified. The terrible words wrenched from his agonized soul, *Why hast Thou forsaken me*, point to mistake and subjective expectation secretly entertained up to the last minute, to a neglect of observation of facts, and to a non-comprehension of the world about and above. His beautiful socialistic teaching would certainly have had added weight, if he had stuck to his carpenter-bench. The same dream-like fervency of world-scorning enthusiasm is seen in Paul's life, and in that of the early Christians and Christian martyrs; it broke into morbid exaggeration in medieval crowd-delusions, in asceticism and in the crusades,—even a children's crusade not seeming absurd; sobered by English stolidity it grasped at something human and practical under Cromwell; and it is now echoed to our shores in the excesses of evangelical Christianity and of the Salvation Army. Psychologically all these things are illustrations of a mental law or habit that is somewhat common, consisting in the luxury or in the utility of emotional exaltation and excess. It is indeed a genuine intoxication, neurologically not much different from that derived from alcoholic stimulation, and in whatever phase presenting itself it is subjective, heedless of facts and of consequences. This power of the mind and especially of crowds, to override obstacle with a saint's indifference, or by a flooding rush of wild emotion, has doubtless served many useful purposes, but it has often also led the hypnotic and frenzied by painful roads to useless death. Surely the heroism of the present and of the future must be a heroism of home-keeping virtues, of reflection, and of quiet persistence of steady emotion throughout a life of loyalty to God and to man. The Kingdom of Heaven can be taken neither by dreaming ecstatics nor by storming warriors.

PROOFS OF THE EXISTENCE OF GOD, logical, ontological, cosmological, and void-o'-logical are for a time amusing, but soon become "stale, flat, and unprofitable." If instinctively one have the belief in his heart, he would better never read the "proofs," and if he has not the belief, these proofs can certainly not supply it. No atheist was ever converted by them, and doubtless many atheists have been made by them. I remember how shocked I was to find that there was no proof of what I already believed, and the discovery was the initial step of examination that for many bitter years led me a sad dance through atheism and pessimism. It finally occurred to me that fact needed no proof. Some funny old metaphysicians, indeed, have reasoned as to the fact of an apple, parcelling out the "data of experience" and the "forms of thought" that go to make up a "sensation," a "percept" and a "concept," until the fruit has faded into "nowhereness," and one feels like a child with watering mouth, cheated out of its apple by some tricksy nonsense of a sleight-o'-hand man. It is precisely the same way with "reality." Metaphysical hair-splitting will not give it to us, and will seemingly rob us of what we had. Facts need no proving. If God is a fact He will doubtless become evident without much help of proofs. This makes me doubtful about the possible usefulness of additional arguments, and yet I feel that to certain minds in a peculiar phase of development the following thought may be of use, as it has been of use to me. It might be called the Booby's Proof. When one is bound hand and foot, incapable even of winking, by the awful bonds of "immutable law," with which the stern atheistical master of our modern mechanic world has harnessed all phenomena, actual, past, or possible, one becomes thankful even for a little glimpse of a free world between the rifts of bandaged eyes. The physicists and the mechanic philosophers have shut God into such an iron mask that we must borrow their own key to get a little glimpse of Him.

A pile of cannon-balls will for all eternity remain as it stands, so long as the foundation on which it rests is not disturbed, so long as the balls are not moved by any external mechanical force, or so long as they do not rust or are not melted by heat. As to the heating and the chemical action, the mystery of their forces is fortunately cleared up by the knowledge that radiation and the compounding of elements, or the dissolution of compounds, are all due to purely mechanical forces. Just so literally as the blow of the suddenly-made gas in the cannon knocks the ball forward, just so exactly does the blow of a striking fellow-atom or of an ether-wave, hit the individual atoms of the cannon-ball and produce the phenomena we call heat and chemical action. Therefore, in the water, in the heap of stones, or in cannon-balls, no other forces than those named mechanical are required to account for all molecular changes. But a bird alights on the topmost cannon-ball with some hairs and dry grass-leaves in its mouth. Here is a new phenomenon; it was neither knocked there, nor was it put there by any mechanical force, but itself utilized all mechanical forces in getting there: it had indeed directional power over mechanic forces of all kinds. In the simplest form of the argument, it had spontaneous motion;· and the wisp of building material in its mouth pointed to purposive action, to a thousand indirect means towards the bringing into existence of other beings like itself, with relations and effects beyond computation. In a word, then, the existence of spontaneous motion, coupled as it always is with directional power over mechanical forces, and with purposive activity, is a fact that cannot by any sane reason be identified with mechanics, or named mechanical. It is absolutely unthinkable that mechanic forces could have produced it. It is a separate, distinct, *sui generis* order of fact, and however limited one may please to think its scope or significance, it fairly, squarely, and indisputably introduces into the universe and into the mind of the young philosopher, the facts of freedom, of intelligence, of design, and of the utilization of, not government by, mechanic forces. Now a universe in which there exists the least degree of these elements is not mechanical, is not " bound by the rigid and unexceptional reign of law,"— and all that. The simplest bit of alga, diatom, or of self-mobile protoplasm, proves God's existence. The shackles and bonds of determinism, that a crude science had woven about every muscle and thought, drop off as if by magic, and the mind looks about at a world limitedly free and of intelligent interest.

It may be said that if the argument is valid, it gives us but a poor God; I answer: 1, that if you don't see and perceive God, no reasoning or proof will enable you to do so; 2, that if you do not wish a God, no argument will give you one; 3, that the argument is more valid, and to a normal fair mind, far more convincing, than any I have heard of; 4, that a poor God is far better than no God. I emphatically contend, however, that this God is by no means a poor one. In the first place He is a living, a real, and a present God, and this cannot be said of the Gods of the old arguments, and of the old religions; and in the second place, study of the actual work of this Divinity, following up the necessary implications of the argument, lead to the evidence of His present and visible richness, and to glimpses of coming greater fulness. The bird and every living thing show mental use of physical atoms and forces; the bird's nest and the reproductive function of every living thing show mentality behind minds, using individual organisms for ulterior purposes,—absolutely contradicting or overriding the "law" of the struggle for existence, the "law" of self-preservation, or whatever other "law" man's prejudice may have devised to hide from his perception the fact of intelligent, purposive, extra-organismal control. Lastly, the creative instinct in man as shown by musical and other arts, the designed production of new fruits and animals, or the improvement of already-existing ones to such a degree as to be of marvellous service—such things point to the living Divinity working to ends the incomparable richness of which are but hinted at in the splendid attainment of present science and art.

Another argument to me of an equal conclusiveness, either unknown or unemployed by those who instruct the Godless how to find Him by the road of ratiocination, is the following: In all purely physical phenomena, the relation of cause and effect goes on producing a series of understandable sequences, all united in clear order, all depending on one another, and all explained by mechanic principles. However often reflected from a thousand successive objects, however transformed into other types of force, a beam of light produces or itself becomes effects, that are of a common kind and comprehensible. But let it strike an eye and it finally results in a "reaction" not comprehensible, and not to be classified in any mechanical or physical category of existence. All physical things and forces are "objective," lie outside of a metaphysical subject that is affected by them through the medium of sensation. All sensation begins in an understandable physical force, and ends in a non-understandable consciousness. The physical stimulus is passive, unimportant comparatively; but the sensitive-plate of mind is active and all-important. However the mechanism of connection between the perceiving consciousness and the perceived thing may be explained or construed, the fact is indisputable that behind the "object" (in which is comprised the ultimate physical end-organs of the brain, the last intermediating agents) there is a reflection or prolongation of the material force over into an order of existence that by no one can be called physical or material. Out from this extra-physical realm there return intelligent and purposive responses. The hyperphysical therefore is a fact, and this hyperphysical is mental, volitional, purposive. The essential identity of all living natural things, and the multiform interdependence and relation of all these hyperphysical existences, point by the most incontrovertible logic to an organismal unity of source and being. Of course no Infinite or Omnipotent is gained by this road, but none such is desirable. We reach, however, a working and actual God, of very satisfying proportions and powers, and we are forever relieved of "rigid law," materialism, determinism, and all that.

Another form of this argument would be this: Classing the ether, the lightest of known substances, with the physical, the physical world consists in one single phenomenon, vibrating or revolving atoms, particles moved by each other according to pure mechanic laws or methods. There is nothing in the "spontaneous" motions of matter except gravitational phenomena. All mechanical or non-biological phenomena are explainable in terms of gravity, which is absolutely a non-mental force. Study and observe for a million years, and nothing else exists "out there," but waves or undulatory movements in predictable regularity. Include the physical parts of spontaneously moved bodies, and from the ether-waves of light, the air-waves of sound, and the molecular or atomic waves of heat, taste, and smell, along the white conducting threads to and including nerve and ultimate cerebral centres,—nothing is discoverable but waves or vibratory motions, impelled by understood and unexceptional mechanic forces. Without importing among these waves mentality, without an illogic reading into them of extra-physical fact, no mentality or extra-physical fact is discoverable. But, mark well that volition, mentality, subject, are parts more indubitably existing than physics, mechanics, and object. To deny these psychic facts is first and *ipso facto* to deny the physical facts, because the knowing of the latter and the asserting their existence, is possible only by means of and in terms of the extra-physical.

CHAPTER III.

INCARNATION.

IT is customary for "the enlightened" to sneer at the "old-fashioned" religious doctrines, and people whose function in life is to sneer, so long as they will not take themselves, may as well take these doctrines as objects for their scoffs as anything else. But any great belief of the human mind must be founded upon some great truth. However much it be modifiable by another truth, it is never an absolute untruth. Primitive and wide-spread beliefs constitute their holders the first philosophers. The function of perception and judgment in the most savage is not so wretched as to cause great bodies of people through centuries to believe the unbelievable. Indeed, untrained thinking and belief is usually truer than the half-trained variety, a little knowledge being proverbially a dangerous thing. A little enlightenment of the sort so fashionable, makes easy the sneer against the popular religious dogma of special incarnation, but no incarnation at all is still more untrue, and neither believer nor scoffer is apt to catch the great truth that the living world, both vegetable and animal, is in its vast entirety a literal and glorious incarnation. There is no perfect religion, nor is there perfect philosophic perception in whomsoever does not zealously believe this. It is the chief and keystone article of the creed of the coming religion. This article in no way denies, indeed, it implies that there are many degrees of incarnation, but it leaves no tiniest living cell or humblest living organism without divine paternity and guidance.

The fact is of course one to be perceived, and no amount of reasoning, again, will give one such perception; but in spiritual as well as in physical failure of vision, artificial aids may be of great service. To psychic myope or astigmatic, then, one may say, "You see nothing but the evidence of adamantine law and blind impersonal forces" in the whole biologic world? How is it that I see not a whit of evidence of either, since to me it is freedom and personality that chiefly characterize every cell and organism of that world?

We should first note that there is spontaneous motion in these beings. All purely physical things are moved only from without, but every law or force of physics is used and dominated by extra-physical volition. The bird dropped in air does not fall. There is everywhere control of the dead material of the inorganic world. Besides this the control is mental or purposive. No mechanical force compels the flower to bloom, the spider to weave its net, the bee to store honey, or the man to build ships. Such things are done in obedience to foresight, fear of hunger, etc., hyperphysical capacities and qualities, surely. Moreover not even "the preservation of self," not any egotism of the lowest or highest order accounts for the actions of these organisms, because an ill-nourished tree will kill itself in the supreme spasm of producing its seed, and "nature" will suck the phosphates out of the bones of the poorly-fed mother for the sake of the unborn child, and if she survive, the same mother later will knowingly rush to death to save her drowning child's life. So there is control of the organism by an intelligence above all organisms. They are never reduced to an automatonism or to a ceaseless slavishness of obedience: the extra-organismal command is always there.

The instant thought of the balking mind is easily met: "Why," it persists, "Why is it necessary to identify the life-principle of organisms and God? Surely the highest wisdom and power displayed in all biological history and science is far short of what we seek in a Supreme Being." I answer first, that the Supreme Being is a credal myth, a speculative fancy, and a useless luxury, both of religious and of philosophic thought, with which as sensible men living in time and space, and governed by fact, we have nothing to do. It was a natural hyperbole of adoring hearts, in large measure true, but in full literalness not true. The real divinity of Whom we thus get knowledge is quite "Supreme" enough, finite and unomnipotent as He is, to almost take Him out of the reach of our love and comprehension, without adding unauthorized attributes, to carry Him farther away from our dazzled power of vision. Then, secondly, the identity is proved, by the unanswerable logic of the maxim, *qui facit per aliam facit per se*. The intermediation of subordinate gods, after the fashion of the gnostics, avails nothing, but only complicates the difficulty. What God does through His agents He is responsible for. Indeed, the largeness of God is really lessened thereby. The theory makes Him more instead of less finite, in being Himself incapable of the details of the great work, and in being necessitated to carry it on by subordinates and vicegerents. I prefer rationally to see, and religiously to love, the one great God, than any under-officer. Any supposition except that of His direct and immediate infilling, lands us at once in a tangle of absurd contention, contradictory theory, and polytheistic mystery, which lead inevitably to the death of clear thinking and of living religion. There is no abiding satisfaction of intellectual honor or heart-hunger except in the unequivocating, unexceptional statement that every living organism of the world is what it is by the literal and constant incarnation of God Himself in this form and manner. He Himself is there, in person and power, its life is His life, its doing is His doing. "Pantheism?"—yes, if you leave out all but an infinitely small fraction of παν as the part with which alone θεος has to do. "The creature becomes automaton and freedom is lost?" O no!

The indestructible basis of this very modified Pantheism, its mechanics, if one may so speak, is as demonstrable as any Euclidean theorem, and consists in the necessity of an extra-physical force for the production of the cell,—the physiological element or unit. The building and nutrition of a cell from its inception to its death is inexplainable by the action of mechanical forces alone.*

* The chemical operations performed by the living cell cannot be imitated in the laboratory or explained by any known chemical laws.—Halliburton, *Handbook of Chemical Physiology and Pathology*, p. 210.

Since this chapter was written the following words from a medical work of the highest practical and scientific character have been accidentally found. They are so apposite that I give them in full, italicizing certain sentences that seem to me especially suggestive.

"It was an epoch-making advance when the old vital forces were dethroned and only physical manifestations were allowed to explain the operations of the organism. The physical methods of research were adopted and the vital processes were placed on a corresponding basis. This was the first step which absolved physiology from its long bondage as a subordinate part of anatomy and elevated it to an independent science. But the fond hopes which were placed on purely physical explanations even up to a few decades ago have since been proved to be unattainable, and the inevitable reaction has set in after we have in vain waited for the solution of all problems by physical science. Even some of the most enthusiastic investigators, who had placed implicit fate in these explanations, now ceased to blindly

Without the instant support of Biologos a cell falls to death as certainly as a dropped stone to the ground. The dead atom falls apart, or rather it comes under the unhindered propulsion of mechanical forces alone, and that

follow this alluring path. Not that there was a reaction to the old vital forces; not that every attempt at an explanation was rejected in despair ; but experimenters became convinced that in many, in fact in nearly all the better known phenomena the physical laws did not suffice to give a clear explanation of the mysterious vital phenomena. *Unfortunately, we are now nearly everywhere compelled to assume a specific yet absolutely unknown activity of the living cell.* This reaction was very beneficial ; it unmasked an apparent knowledge and brought us nearer to a true understanding of nature ; and, even if we must finally admit a mechanical basis, yet we are still infinitely remote from the goal of all natural science. That we can only reach this goal by extending our knowledge of the vital phenomena in the individual cells is the advance which has resulted from the reaction against purely physical speculations. The same conceptions which elevated physiology to an independent science would merely have converted it into physics and chemistry as applied to vital phenomena. Now, however, its character as an independent science is forever preserved.

"The functions of the stomach consist mainly of secretion, absorption, and motion. It was once thought that the activity of the glands could be explained by the purely mechanical processes of filtration and diffusion. The chemical and physical changes in the blood circulating about the glands, of which the physical were regulated by the nerves, seemed sufficient to explain why the secretion of one and the same gland may vary in strength and composition.

"Although Johannes Müller had long ago called attention to the specific activity of the glandular cells, yet only recently was it *positively demonstrated that the mechanical processes of filtration and diffusion do not suffice to explain secretion, and that we must accept the existence of a peculiar activity of the cells.* [Foot-note. Ewald, *Klinik*, etc., I. Theil, 3. Auflage, S. 61 und 208 *et seq.*] Nerves may regulate this cellular activity, yet secretion is unquestionably possible without them, and in this respect the animal tissues do not differ from the vegetable, which have glands but no nerves.

" In the process of absorption the specific activity of the individual cells becomes even more obvious. Here, contrary to physical laws, some substances are taken up, while others are rejected. The lymph-cells have been observed to wander to the surface of the intestinal mucous membrane, and organism is no more. Couple with this the fact that the humblest living cell always shows mentality, and the physiological evidence is clear that divinity is more intimately and more vitally present than any deductive faith could

there incorporate drops of fat ; they then creep back, even into the lacteals, where they give up these particles of fat. In the face of such occurrences, which seem to play an important part in absorption, how can we think of purely mechanical explanations? At all events, *in the processes of absorption peculiar functions of the living cells must coexist with filtration and diffusion.*

"The conditions are no more favorable in the motor function. I disregard entirely the fact that what occurs in a muscle during contraction is as incomprehensible as what constitutes innervation in a nerve. But the dependence of the contraction upon the nervous impulse, and the invariable result of this impulse, namely, a shortening of the muscle, were formerly regarded as a general and, in a certain sense, physical law. Indeed, for striped muscle it would be difficult to find an exception to this law, if we do not include the direct stimulation of the muscle which can only occur in an abnormal way. The striped muscle-fibre is always at rest till an impulse reaches it through its nerve ; the result of this impulse is always a contraction, be it a jerk or tetanus. The apparent exception that the heart continues to beat even after all its nerves have been divided, was explained by assuming that the impulses may arrive in the heart itself in the ganglion-cells and that these impulses are transmitted to the cardiac-muscle fibres through the intracardiac nerves. It was, however, discovered that *sections of the heart which positively contained no ganglion-cells continued to beat rhythmically.* The greatest difficulty of maintaining the law of the dependence of muscular contraction upon nervous impulses is encountered in the unstriated muscles. Here we not alone observe movements which are independent of any nervous influence, as for example in the ureter, but we are not even able in every instance to prove that the result of the nervous impulse is a contraction of the muscle. Thus irritation of the vaso-dilator nerves causes the arterioles to relax, and as, for many reasons, we cannot explain this by the longitudinal fibres we are compelled to assume the paradox that the circular fibres lengthen upon irritation. We must therefore admit that, with the possible exception of the striated mucles, the above law does not always operate, and that consequently *the muscles may both make spontaneous movements, and may also lengthen upon stimulation.*

"These preliminary remarks will enable us to comprehend more readily the unpleasant fact that we know very little about the secretion, absorption, have conceived. With this conception of incarnation, of the immanent in-living, up-bearing, through-thrilling Father of Life, struggling under difficulties with rebellious and dead material, comes the perfect solution of most of the mysteries that have always and so sorely grieved the heart of man, that have made the course of history so puzzling, the records of biological history so gloomy, and our way in time so aimless and hopeless. The reason of "evolution," or progressive process, becomes clear; the question of the origin and nature of evil, that like a terrible Sphinx stood in every path, devouring whoever could not solve her insolvable riddle, is now smilingly answered by the Sphinx herself. Duty is no longer a blind, stern force pushing us nowhither through darkened, roadless woods, but a loving pleasure-walk to distant hill-top outlook, with peace in one's heart and happiness ever at one's side; the shudder and the gloom of death itself are forgotten, and a rational immortality takes the place of the sensual or *ennui*-filled heaven of the objectless after-life of childish faiths.

and motility of the stomach. The experiments are very frequently contradictory; many contain conditions which, upon closer examination, preclude a uniform result. It is evident that *the study of the organ has been undertaken with too many physical propositions, whereas here, as in the digestive tract, biological laws are more important.* It seems that the more highly vegetative the functions of an organ are, the more does its activity depend upon its own cells, and the less upon the nervous system. In fact, could we remove every nervous element, nerve-fibres as well as ganglia, from the walls of the stomach without injuring other tissues, it would secrete, absorb, and contract quite well. One may ask, why, then, all these nerve fibres which enter the stomach? For the same reasons that nerves go to the automatic heart—to connect it with the rest of the body. On the one hand, the stomach has these connections with the central nervous system to fulfill the demands of the other parts of the body; and, on the other, to enable the entire organism to take cognizance of its condition."—*Diseases of the Stomach*, Ewald, p. 363.

The Mechanics of Incarnation are at present locked up in the mystery of the living cell, or more accurately of the living somacule of bioplasm. The secret of the method of God's entrance into and use of matter lies here. In inorganic chemical compounds there is the basis or hint of the system-formation upon which Biologos builds the living molecule and cell. The solution of the mystery of mechanism of the living cell is partially dependent upon the solution of the mystery of inorganic chemical union. Both are at present somewhat hidden from us, though both will doubtless be solved by progress in microscopy, spectrology, and thermal chemistry. In part we are coming to understand the mechanism of the inorganic chemical compound. It is probably entirely explainable by the simple law of gravity and momentum, and in precisely the same way as the solar system is explainable. The application of Newton's laws to the inorganic chemical molecule may explain all interactions of its constituent atoms, and the qualities developed by their union.

A system, *i.e.*, a compound, is only formed when the rate of the periodic vibrations of its constituent atoms is such that they revolve about, or in relation to each other, indifferent to or resistent of the weaker gravitational influence of neighboring atoms with inharmonic or aimless vibration. A system is formed when individual and *aimless swing* is transformed into *periodic revolution*, in conjunction with another or other atoms. In the mechanical mixture forming our air each element has a periodic vibration (or specific weight) out of mathematical proportion to that of the other, and it swings aimless, under the blows of colliding atoms. In the formation of water, however, a genuine chemical union or system-grouping takes place under certain conditions, whereby two atoms of hydrogen and one of oxygen are caught in harmonic phase, and mechanical aimless vibration becomes interacting revolution, with closer condensation and resultant density of product, and with just that largeness of free path of revolution which above 32° F. results in the condition of a liquid, and at other temperatures, in that of a gas or of a solid.

Viewed as a unit the solar system is a closed system, and it is truer to say that the sun and any planet revolve about each other rather than that the planets revolve about a fixed center. The solar system typifies in large the complex, inorganic molecule. In fact "matter," though to a lesser degree, is like the solar system "open." The essential difference between the solar and the chemical molecule is that the latter can be crowded upon by adjacent atoms or systems, and thus changed in characteristic; can be relieved of external pressure by greater distance of adjacent systems, often with consequent results; or it component atoms may be struck by the blows of in coming ether-waves, and thus swung out of the system with resultant disruption of the compound. In the sol: system nothing of this kind takes place. There is r other solar system so near as to crowd or relieve fro. pressure—even Neptune being so far away from the neare external neighbor as to be comparatively uninfluenced t. it—and there is no interplanetary fluid except the ether, t! blows or waves of which are so infinitesimally small th they can only exert mechanical pressure or tension upo atoms as small as themselves, and in vibratory period h: monically related to them. Now, the supposable exhibiti of purposiveness, or of mental control of the solar-syst could only be conceived as taking place through an *extra* will coming into it, with ability to lessen or to increase the distances from each other of the globes, to draw in from without other complex systems and exhaust them of their force or heat, thus breaking up their combinations of vibrations and stealing the vehemence of their swing‑ ing, and stilling them down. In physiological language we call this on its constructive side, anabolism or nutrition, and its reverse aspect is named katabolism or retrograde metamorphosis. If there were a giant deity capable of do‑ ing this, the solar system would become what we call living, *i. e.*, it would show the functions of a living molecule or cell. The difference to be noted consists in the number and complexity of the constituent elements, the solar system being composed of a limited number of planets, some with simple subordinate systems, asteroids, and comets, like the inorganic compound, whilst the living molecule of protoplasm is made up of an unknown num‑ ber of constituents—1,000 or 2,000—(the ignorance a re‑ proach to chemistry), grouped together in a bewildering complexity of intricate and involved sub-systems with marvellous dependencies and inter-relations.

Now, although no such evidence of *ab extra* influence or of volitional utilization of the solar-system molecule exists, just such evidence and use is shown by the actual miracle of the formation and the function of the physi‑ ological molecule, or somacule * and upon the fact is based the existence of every living organism. The entire mechanism of incarnation depends and hinges upon this secret. Biologos gets control of matter in this single and

---

* The molecule is the smallest quantity of a substance representative of its qualities, capable of existing in a free state ; it is the typical unit of the compound, and should be limited to inorganic substances. The somacule is the organic molecule, and the cell an organized system of somacules.

only way, and the entire drama of organic life on the globe is dependent upon this mechanism of molecular formation and control. This is the essential principle of biologic philosophy. Firmly grasped and comprehended, it will be seen that from it radiate lines of light like the karyokinetic figures of cell-division from the nucleus * into all the darkness of religion and philosophy. Cytology, the secret of physiological chemistry, is the secret of the mechanics of incarnation, but it is also very largely the secret of the greatest problems of philosophy, of religion, of ethics, and even of esthetics.

I am well aware that the *streng wissenschaftlichen* sticklers will aver that any such interference with molecular physics contradicts the law of the conservation of energy. Kinetics, they say, can know no exception to the law that "in a closed system the sum of energy is the equal of the sum of the specific heat units of its elements." The least directional energy, it is said, brought to bear upon a process, seems to imply the importation within it of an *increase* of energy, and of that our finest analysts with their finest bolometers can detect no evidence. This is as much as to say, "you can't get more meat out of the egg than there is in the shell." The contradiction is only seeming, and is another of the many ways men take to cheat themselves with words, instead of trusting to first-hand observation of facts.

Assuredly no bolometer will ever detect it; nor will any physical thermometer ever do so. The only weigh-

---

* "The cytasters and radiating lines in the protoplasm around the poles of the spindle of a dividing cell remind one forcibly of the effect produced by placing a magnet in the midst of some iron filings, the radiating position of the metallic fragments around the poles of the magnet indicating the direction of the lines of force."

ing machine delicate enough to react to this stimulus is the pondering instrument that lies behind human perception. When the miscroscope enables us to see atoms, and when ocular muscles are nimble enough to move synchronously with them—some several hundred millions of millions of motions per second—then the bolometer may be invented that can measure the force of God's finger in the somacule and cell. In the meantime we wait, reckoning it better to believe our spiritual eyes than to trust to a crude physical law, physicists having so often proved themselves mistaken in their judgment as to what is "unexceptional law." The criticism to us says only that life-force has not yet been correlated with physical forces. Whatever form of mental picture we make of life-force, it must be deemed of infinitely greater tenuity and subtlety than etherial forces, and it is only yesterday that science has told us anything of the ether. One hundred years ago it was not necessary that we should understand spectrum-analysis to believe in the beautiful sunshine. To-day the science of biophysics is not born, but we can quite as truthfully believe in Life—because Life we are, and Life is the fact that, of consummate interest, is everywhere most real and evident. One of the severest tasks of philosophic thought is to hold fast the truth already gained, while grasping and allowing validity to a higher and contradictory truth, contradictory only in temporary seeming, because all chromatic half-truths are by the achromatic lens of honest reason finally brought to focus in a higher unity of pure white light.

THE UNITY OF ALL ORGANIC LIFE is rarely acknowledged, and even by those acknowledging it, its remote consequences are little appreciated. Almost every day one will find writings that separate man and "Nature," and, by a previous jumble of inorganic and living "Nature," make man the enemy of all Nature, each following different and cross purposes. To confuse under the term Nature two such different orders of phenomena as the living and the non-living, may pass in Sunday-school books, but there is no school-boy now-a-days that should not understand that man is part of the great order of biological phenomena. Any good text-book of botany will prove that the protoplasm of the lowest vegetable organism presents all the qualities of the protoplasm of the highest animal organism. It is needless now to emphasize the admitted fact that no hard and fast line exists between the plant and the animal, many species being in the neutral bounding territory, and that in some degree all plants possess many of the intelligent purposive functions, whilst in moderate degree many possess all of the essential characteristics of animal life. The difference that exists in the fact of the power of the plant to eliminate living cells from mineral or inorganic material alone, whilst the animal is dependent upon the vegetable world for its supply of this prepared food, really argues against any essential difference between the two types: it is simply a differentiation of function, and argues for unity, since it shows an interdependence and close vital relationship. Every agricultural or botanical chemist also knows that the vegetable kingdom is also dependent upon animal functions and activities for its stability, so that the relationship is one of mutuality in ways more subtle than the striking example known to all, of the complemental service of bee and flower.

The extremely complex protoplasmic molecule of the animal is built upon the moderately complex molecule formed by the plant: this is the striking and highly suggestive arrangement upon which depends the existence of animal and human life on the globe. Except in some possible rare instances the animal, if it ever had it, has now lost the power of constructing its unit of physiological activity, the cell, out of crude inorganic materials. This, our utter dependence upon our humble "little brother," furnishes an intelligent explanation of one of the greatest mysteries: the apparently reckless, useless, wasteful luxuriance, the fighting, clinging pertinacity, and the infinitely ingenious device, whereby millions of forms of plant-life find means to penetrate and live upon every possible inch of the world's surface. It might seem that the omnipotence of an infinite spendthrift were intoxicatedly scattering the riches of eternity in heedless profusion in the few hours and acres of our little planet. But if we could look with the large vision of Biologos, I doubt not we should reckon it all as the most economic wisdom.

Should we conceitedly question the foresight that for millions of years transformed dead atoms into living trees and ferns, and these into reservoirs of force, our coal-measures, upon which our civilization is at present dependent? Always distrust the wisdom that deems itself wiser than the wisdom whence it sprang. Harmful, useless weeds and noxious plants and trees!—I doubt if there be one such. The deeper the insight the less the criticism of God, and the more the old-fashioned faith comes back that every living thing "has its use," present or possible. And when all criticism is allowed there remain the indubitable facts that all animal life, that of man most of all, depends upon all vegetable life, and that, in the thousand-fold danger to which any one type or all types are always subject, from volcano, recurrent glacial period, submergence of islands or continents, vegetable disease, parasites, or destroyers of exhaustless variety, upon any single order or family—nay, upon any single plant, that by limitless pertinacity and ingenuity had saved itself in some wreck or lethal change, may depend the related existence of other forms of life, even the human; its "pluck" may rescue from absolute destruction the results of the combined effort of production of past ages. In any such general death or destruction one rescued seed or plantlet may be of incalculable service to any or all animals, besides giving Biologos a foothold from whence to re-people a continent with that type, or from it to re-develop new species. The adaptability of any one type to change and variability heightens its possible value, because it is always easier for Biologos to develop one type into a different, than to create a new one immediately. There is more than a pathetic beauty in such instances as that of the little hairy wood spurge that before the Thames and Seine were disunited by the English Channel produced by the drop of England into the ocean, had crept along the earth from the warm South to England, and there for thousands of years has sustained a precarious existence in one or two sheltered nooks or by some warm spring, while its nearest sisters frozen out of the north have had to retreat to the Mediterranean shores again. Upon several American mountains there are yet left glacial butterflies, stranded by receding ice-tides some eighty thousand years ago, still brightening bleak solitudes with beauteous life; they may be thought of as awaiting their cousins when the returning glacial period may bring them again, although ages hence other catastrophies we know not of may endow them with a superlative and practical use to our descendants.

The vanity of man makes him assume a competence and a right to judge and condemn the whole order of things of which he is a part and out of which he has sprung. "Ah!" we always hear. "Ah! If God had but been possessed of my wisdom and righteousness. What waste, what blundering! What egregious follies mark the evolution of life and blacken the sorry farce of history! Nature is a stern and relentless blind mother, and her cruelties are as heartless as her follies are stupid." Much such nonsense one is forced to read and hear every day.

One feels like asking these silly beings if they were produced outside of "Nature." Did some other faultless "Nature" beget them and their high critical powers? Whence do they draw the sources of their being, and the mental outfitting they so admirably use? At least they are guests of Nature now, however daintily exotic their supposed origin. They are living off her bounty,—and like filthy lunatics they abuse and vilify the kind host that nourishes them. We may never safely lose sight of the divinely suggestive fact that if *we* have anything admirable in our human nature, if reason, love, honor, character, are ideals or actualities of our life, if benefaction, order, freedom, are our acquirement or seeking, if science, poetry, art, music, are our creations and delight, they are primarily the handiwork of Him who made us. We did not certainly make ourselves. These things cannot be conceived except as the blooming and fruitage of a tree planted and nourished not by men's hands, for they all sprang and still gush not from planned effort of our conscious seeking, but from the still deeps of being below and behind our personal consciousness.

The destruction of the interdependence of one class of animal organisms upon another is a fact of which we have as yet gained hardly a hint. Some writers have touched upon it, and catch glimpses of the sad disturbance of power such interference as our brutal hunting causes, in destroying the very agents that keep our living world healthful. The otter takes the salmon blinded by fungoid disease, and birds of prey capture the grouse weakened by disease that would infect the whole family. The unavoidable conclusion of such a careful student as Darwin is that in the struggle for existence the vigorous, the healthy, and the happy survive and multiply. Among the most pernicious of man's actions in this respect is the wanton destruction of forests. The perturbations, the echoing and re-echoing effects of desertizing the globe, in climatic and hygrometric changes, in the pernicious upsetting of a long and laborously-established harmony of vegetable, animal, and human life, are beyond finite calculation. One rises from the perusal of such a book as Marsh's *The Earth as Modified by Human Action*, with a pained sense as of a horde of Northmen destroying or putting to Atilla-use a gallery of Greek marbles. How can the Infinitely Patient and Ingenious Healer ever curb and enlighten man and repair the damages of his invasion?

But, just when hopelessness seems the only fruit to gather, there is growing a God-planted "seedling": Forestry is becoming a science, and humanized, or divinized Science is dextrously setting herself to discover, to save, and to repair. There is in our modern love of the living world, and unconsciously beneath the beautiful zeal of science, an unseen wisdom working to heal the hurt, and undo the injury of the horrible years when the animal was struggling to a yet-unreached civilization, and harming all about him with the insanities of a sated but undigested animalism, and a tasted but unattained humanization. " Man opposed to nature?" "Wresting from her her sullenly yielded secrets?" "Fighting her invasions and enmity?" —Ah, me! How blind themselves are these philosophers! The purer and the clearer eye recognizes that all life is interdependent and a unit; that the greatest king is beholden to the meanest diatom, and to the humblest weed of unexplored regions; that fungus, tree, animal, and unhumanized savage, are all dumbly longing for the love and the guidance of their rational, seeing brother, man. Biologos will hold them with His soft slavery, He will keep them, and they will wait for the Coming King, governed in the meantime by His agents, the instincts and the "struggles for existence," the subtle balances of power, the curbs of hypertrophies and stimuli of hypotrophies,— His only but effective means,—until He shall have so educated and deified man, until He shall have so filled his heart with helpful love, as to operate through him by more effective and direct means to perfect life upon the globe. The great world of living things awaits "the coming of the Lord," the deputy of God, man, to give it freedom, to bring it release and guidance, to educe soul out of sense, and to teach the inexhaustible beauty of love. Poor half-savage and shipwrecked folk, we have clutched the island of Time on which we have been washed, and seek to crush, and enslave, and devour the little brothers stranded here before us, and who in the common misfortune that has overcome us all, would better deserve, and more infinitely repay, our sympathy and our unselfish help. "Thou shalt love God with all thy heart," acquires a divine significance when we come to recognize the real "God," and to know all of "our neighbors." The humanization of our friend the dog, the domestication and improvement of fruits and animals, the miraculous *response* of all living things, heretofore loved and cultivated by man, give hint of the waiting readiness of the living world, and of the wonderful results that will follow when we learn to love and to help all instead of hating, or at best selfishly using all. Happy he who shall be the new Christ, the re-arisen and extended Christ of this new religion, and shall both point and lead the way to this coming Kingdom of Heaven in all the Earth!

CHAPTER IV.

CYTOLOGY.*

CYTOLOGY IS THEOLOGY, because a perfect knowledge of what takes place in the cell, a knowledge of what causes the changes, and of the ultimate effects of these changes, would let light into the whole "before and after." I can only give a few of the glimpses of the implications and of the consequences that I have caught, but below the horizon's edge it is bright with a light I fain would travel to see.

Ὁ λόγος σάρξ ἐγένετο: The Word became Flesh. So said religion most truly, and yet religion has never shown the faintest curiosity as to how the Logos made unto itself a beauteous instrument of flesh. And the bitter irony is heightened by the sad spectacle of a Godless Science blunderingly seeking to discover the method, whilst at the same time, far from working herself, and far from showing any interested encouragement, Religion stands by and affects to scoff at the blind worker. Such are the spectacles that meet the philosophic eye everywhere. We are all working to hidden ends by mistaken motives, and the light will come when the divinely beneficent smile reveals the reward of unexpected success greater than would have been that we delusively sought. The honor and the success will be in proportion as our work has been guided by a self-effacing loyalty to truth.

* Κυτος, cell ; λογος, science, the science of the cell and of Cell-Life.

The cell is God's που στω for moving the world. Its nucleus is His millionfold point of contact with matte.. Back from every cell-nucleus run the lines of unity, meeting in the focus of His personal control. Language fails us when we try to picture to ourselves or to others this almost incomprehensible fact. Perhaps it would b. nearer truth to say that Spirit, which is outside the limits of space and time, finds, or rather makes, the cell His universal throne, so that He Himself in very literalness is the life of each of the innumerable tiny elements tha' make up the substance of plants and animals. In this fac we get the vivid demonstration of many things entirel* extra-cellular. For instance, the infinite detail of Hı work. The atoms of every cell could not place ther selves as they are in the cell, they could not build the ce as it any instant is, or for a moment keep it in function Constantly rebuilding and upholding, an extra-cellul control of every cell is evident.

But every cell differs from every other cell. Not on¹ does each cell of all of the myriad classes of tissuc have a typical likeness to those of its kind, and a utter unlikeness to those of all other kinds, whethc of liver, kidney, parotid, contractile, constructive, (osseous, but it is only our finiteness that keeps us fro. realizing that the individual cells of each class diffc from each similar associate. The typical protoplasm: molecule of each planet must differ in general chemic. formula from that of every other planet, that of the veg table from that of the animal, every class of each, in larg

* Despite such oft-repeated nonsense as this : "The movements of mat and the phenomena of mind, are separated by a fathomless abyss. Betw the one and the other there is a great gulf fixed, and neither can matter upon or induce changes in mind, nor can mind act on or induce changes matter."—MERCIER, *Sanity and Insanity.*

ways, from that of every other class, and so on down to each individual, and to each cell of every individual. In men, we easily detect racial body-odors, and the dog can instantly differentiate the odor left by his master's foot from that of every other of the fifteen hundred millions of men in the world. I have alluded to proofs of God's finiteness, but here of course the mind is crushed with wonder at this proof of His practical infinity.

The mathematician can hardly estimate, and the mind can in no way conceive the multi-millions of cells of any one organ of the many that make up our body, or the grass-blade at our feet. It is said that about five million blood-cells die with every breath. The nimblest imagination shrinks and shudders as it peers out through the stretches of space and time filled and palpitant with silent inexhaustible life-power.

With all the repeated life-time labors of great minds on single subjects of study, on single organs of any organism, there is exhaustless mystery beyond. We know the ultimate secrets of no one organism or of any one of its organs. In rapt astonishment we can only catch promises of glimpses and glimpses of promises of the subtle devices, the delicate mechanisms, the out-worked intricacies of construction and adjustment, the far-sighted vision and forefending dexterity that characterize even the simplest organism, and that appal us when we correlate each with all. Linked each to other, intermediate and inexplained result merging itself into cause, kaleidoscopic change bound to ever foreseen purpose, purpose linking itself to purpose, losing and finding itself in higher unities, —the mind is able certainly to know that the living Logos is in all, but it has certainly not seen the limits of that metaphysical mentality.

And in this way we could go on envisaging phase after phase of the many aspects that spring into view as we clearly realize the indirect inferences derivable from cell-incarnation and cell-philosophy.

But certain negative considerations also rise, and warm emotion is suddenly chilled into its isomer, reason. Fate that as servant may bow, but as slave will never bend,—fate is plainly seen in the simple fact that only by the mechanics of the cell could God compass his ends. Matter could only be conquered by the awful indirection of cellular physiology. In other words, molar or mechanical energy (with all correlate forces and distant ends) is directly impossible to Biologos; He can only reach mechanical and meta-mechanical results by the massed effects of billions of separate but conjointly-working cell-functions. The nearest metaphor, or likeness, is that of light. Repetitive billions of ether-waves are required even to beget the simplest sensation of light, but only combined multiples of these numbers can finally evoke the mite of mechanical force shown in Crookes' radiometer, whilst calculation of the waves locked up in a pound of coal is quite beyond finite mathematics. Just such cost of indirection is asked by matter as the price of niggard consent to serve the Lord of Life. God accepted the condition, and the living world stands before us. The fact shows the most decided lack of absolute omnipotence, but, from the other point of view, it also shows the fulness of relative omnipotence.

It is this complete recognition of the labor, the ingenuity, and the love, the infinite patience, and the indirection with which Biologos has circumvented the poverty and obstinacy of matter, and turned defeat into victory, that enables us to sympathize and love Him. Because His work is our work. Throughout our life we are but doing, in our way and limited sphere, exactly what He is doing. We can, indeed, do nothing else. To maintain a perfect incarnation, to conquer, control, and utilize matter, to make spirit speak through physic—this is the work at which life is set, whether the life be that of God or of ourselves. The sacrifices He has been compelled to suffer to compass His object make Him comprehensible and endear Him to us, and these same sacrifices constitute the mysteries of evil, of imperfection, of disease, and of death. Sacrifice is always the key to mystery, and just as suffering brings clearness and modesty and chastening to man, revealing many human mysteries, so God's sacrifices constitute the explanation of most divine and cosmic mysteries.

It is worth noting that our ignorance of matter is, in strict analysis, far greater than our ignorance of God. In the ultimate of self we have direct knowledge or perception of "the ultimate reality that is behind the veil of appearance,"* whilst all our knowledge of matter is mediate, second-hand, and as yet indefinite. The reality is what we do know, and the appearance is as yet the unknown.

All biological facts and truths center and end in questions of nutrition; and all questions of nutrition are solely questions of nutrition of the physiological unit, the cell. The cell, it must be again repeated, is the only agent and tool that Biologos has. Organogenesis is solely the massing together of some billions of independent tools, of which He has the use and control, to educe therefrom mechanical results, the mechanical results or organ-

---

* "We are not permitted to know,—nay, are not permitted even to conceive that reality that is behind the veil of appearance."—SPENCER.

work being merely the summarization of the individual labors of all the cells. The organ is the mere unition of the work of all the cell-elements. In botany and in zoölogy, cellular physiology is the ultimate science to which all other physiologies minister, and in which they end. Indeed, physiology as ordinarily taught and understood in the schools, is only the mechanics of the subject, only the consideration of a few of the crudest end-results. The child with the most rudimentary philosophical bent suspects this, but the questioning is at once hushed by the teacher, who by his "scientific training" prevents any suspicion of ultimate reasons, final causes, root questions, or teleological suspicions, to arise in his own mind or in the minds of his pupils. To the healthy intellect every question of physiology leads by the most natural and inevitable logic to teleology; to the "enlightened scientist" it too often leads to intolerable blankness and self-stultification. The crude end-results that common physiological teaching is satisfied to consider, all lead to cellular physiology, and all cellular function leads as inevitably to extra-cellular life and control, and a multitude of metaphysical facts. The bones are to give the body stability and uprightness, to attach muscles to,— whereby motility is assured; the muscles are the agents directly mediating self-motility; the visceral organs are to elaborate nutritive elements for the feeding of all parts; the vascular system is to carry the food to and from all parts, the heart pumping, the lungs oxygenating; and, lastly, the nervous system is to control all parts and make all functions harmonic and unitary. *Mutatis mutandis,* the same "explanations" of the same functions in the plant organism are held satisfactory.

Now, all this is just as sensible as to explain a

steamship without a thought as to the sheets and bolts of steel on which all depends; without a word as to the coal or steam; the intelligence that created and guides the huge mechanism; or the purposes for which it is to move about the ocean-world. The cells and bioplastic units of the living organism, the combined individual life and workings of which constitute the life-mechanism of the whole, are not spoken of, neither the force that proceeds from each cell-nucleus, nor the mentality displayed in the creation and working of the unconscious machine, or in its conscious guiding will. If, finally, object and purpose of the whole mechanism is inquired about, the poor little dazed philosopher is told by the catechism of the religionists that the object of life is to glorify God and enjoy Him forever, or by the catechism of the scientists that agnosticism is the end of all thinking, and "evolution" the purpose of all existence. No wonder that the stunned curiosity silently concludes that the theologic answer is wholly beyond comprehension and practice, and that the scientific answer is blank pessimism. The resultant "line of least resistance" is that indicated by the indubitable pointing of the lowest senses, and in this way self-preservation and self-gratification necessarily (and rightly, too) become the pursued ideals. Religion and science have both united in the noble calling of making atheists and sensualists out of their pupils, who, in graduating from either school, fling their diplomas with disappointment and disgust into the dying fire of youthful ideals and enthusiasms, and plunge out into the whirl of commercialism and self-seeking. The best and shrewdest of these pupils attain "comfort" and philistinic "respectability"; the worst and the stupidest (with a considerable minority of the unbalanced, unselfish, and noble) drift into galling poverty, sensuality, criminality, and multiform wretchedness. But all alike are plebeian, coarse, without poetry, religion, or ideal—blown by the winds of passion into a future that is dark and hopeless, without rudder of intellectual stability or compass of guiding duty. The atoning and righting civilization of the future will have bitter thanks for the irreligious religion and for the unscientific science of to-day.

Every subordinate organ or part of an organism points otherwhere for its explanation. All parts pertaining to form and motion are to aid in the satisfaction of desires, usually of hunger and regeneration; all digestive and vascular functions are to supply food to other parts; all nervous structures and organs of special senses, to help all other parts,—each is slave to other, and finds no end in itself. But it is to be noted that all organs whose function is to secure the nutrition of others, at the same time effect their own nutrition; that all organs are made up of the same elements, that is of cells slightly differentiated for their special work; and that every living cell, vegetable or animal, displays all the essential functions that are shown by the organ or the organism, or by any other organism. More striking degrees appear by the combined action of many differentiated cells, but the careful eye finds them all in the cell seeking expression and power. Now this is equivalent to saying that the great Biologos, the living God Himself, is present in the cell in full mentality, purpose, and personality, lacking only adequate means or power to more fully reveal Himself and to effectualize His aims, and moreover plainly taking such steps toward this effectualization. The discerning eye can meet divinity face to face in the ameba and its pseudopods, or in the white-blood corpuscle, and can as genuinely and often as justifiably worship Him there as in the temple or the cathedral.

To again repeat: the chief and absorbing function of nearly all but one set of organs is that comprehended under the term nutrition. The millionfold devices of plant and animal for securing food, and for protection against the antagonisms and enemies of life, animate and inanimate, against cold, and against the "running down" of the clock-mechanism,—all are but various phases of cell-nutrition. It is the hunger of the cell that has created the corporeal instruments, and the corporeal unity itself, the more surely to bring about cellular feeding. Of course a not-forgotten and still deeper "final cause" lies behind all hunger and all nutrition,—but of that just now it is not question. The science of microscopy is only an indirect assertion and proof of this fact.

But there is one set of organs, which, together with the emotions that lie behind them, yield no obedience to other organs, are servant to no others, but imperiously demand service of all. To the savage mind these organs were the most indubitable proofs of a mysterious outside deity, whose symbols were most serious and whose worship most sincere, whilst to the profoundest philosopher, they now none the less flash upon the perception, the gleaming evidence of extra-organismal control and of after-coming revelations. After the biologist's nutritional studies, hardly second in importance are those of the reproductive function. The two departments comprise quite all of his work. The primary significance of the one is a hyperphysical Logos-life gaining self-expression and control of matter through the mechanism of the cell; the significance of the other is the perpetuation and extension of that control, and of that revelation, in advance of foreseen failure, by the mechanism of reproduction. There is a subtle but profound significance,—indeed all the significance it has is here,—in the decking with flowers of young-womanly beauty in blooming years or in marriage-days. The flower perfumed with God, the girl his divine exotic, each flooded with unearthly benediction and crowned with His halo of beauty, both alike bespeak the wonder of His promise and the reward of its fulfilling, when we render loving obedience to His command. "Eye hath not seen nor ear heard, neither have entered into the heart of man the things which God hath prepared for them that love him."

The truth that CYTOLOGY IS MEDICINE is illustrated by a number of facts, chief of which of course, is the firm establishment and general scientific acknowledgment of the cellular pathology of Virchow. As all physiology is primarily and essentially the physiology of the cell, so all pathology is necessarily essentially the pathology of the cell. All scientific physicians are conversant with Virchow's great work and of the development of the doctrine, and its exemplification by every new medical discovery, It is therefore useless to further emphasize or detail it. As the body is composed entirely of its cells, it is, indeed, a mere axiom, that disease must be disease of its cells. As physiological chemistry learns the facts of cellular nutrition, it will become more and more clear how the cell's vitality is lessened or interfered with by improper nutrition and microbic enemies. Slowly the prevention of disease is coming to be recognized as the great problem and the established fact. Already knowledge has outrun practice, because if the known truths of preventive medicine were applied, the present death-rate would be reduced by one half. The ultimate philosophical lesson of physiology and pathology is that Biologos is the great Father of physiological chemistry; that our little child-like glimpses of the science is the lisping a b c, of His cosmic science, begun with the first speck of bioplasm that He created, and to-day active in every cell of every living thing. He in person is present in the cell; all physiology is healthy nutrition of His cells and all disease is their disturbed nutrition. All curative medicine is taking away enemies, faulty or disturbing elements, and supplying healthy ones. No bungling medication ever did aught in disease but permit the intelligence and healing life behind and in the cell to cure the cell. The removal of a few obstacles and the supplying of a little proper food, constitute the intelligent physician's task. Biologos in His own unknown way does the work of cure whilst we stand almost idly by, our little wisdom largely consisting in not interfering and in not complicating His struggles.

To the understanding of the world the subject of temperature is of more interest and importance than all the disquisitions of speculative philosophy from Plato to Lotze. But not one great philosopher ever thought of, or mentioned the idea. It was the chief element of God's first great labor, and it is to-day the burdensome problem and exhausting task of an ingenuity and care wholly beyond comprehension. At present the balance of temperature established in man is so extraordinarily delicate and fine that the functions of the most complexly organized centers that intermediate the highest vital and mental processes are liable to fluctuation and danger from the most subtle and unperceived causes. This uniformity and exquisiteness of heat-balance has been established by making the substance of the highest functional cells of an amazing complexity of structure, wherein the change of a degree or two either chills the metabolic process below the most perfect working, or heats it toward dissolution. Our knowledge, our thermometers, and our tests are at present as incapable of proving and putting this in scientific language, as a Hottentot is capable of properly operating and explaining a magnetometer. We do not even know the names of the hundreds of complex subordinate substances, each with its own delicately-adjusted melting-point and qualities, each held to other by the slightest bonds of intra-molecular vibration, united together in subtle couple of intricate balance and affinity,—all held in the most unstable equilibrium, and utilized by the great cell-chemist for definite but by us hardly-guessed purposes. We are so far from definite chemical knowledge that we cannot tell within hundreds, perhaps within thousands of atoms, the sum-total of ultimate elements composing the typical protoplasmic molecule. In a patient a rise of temperature above the normal shows, of course, increased cell-activity and increased danger of the more violent atomic and molecular vibration disrupting the delicate bonds that hold the more unstable subordinates in subjection. Cytolysis results. If the engine is put to too high a rate of speed the centrifugal forces overcome the centripetal and there is death from exhaustion and dissolution. Unless the cell-mechanism is of exceptional stability, fever, or abnormal heat, which are names for increased vibrational or cell-activity, very soon gives evidence of an excess of the centrifugal forces beyond the control of the cell by its master, and molecular, followed of course by somatic, death, ensues.*

* Of the total heat evolved in the body, about 7 per cent. is used in external mechanical work; of the remainder, four fifths is radiated through the skin, and the rest by the lungs and excreta.

But it is of exceptional interest to note that subnormal temperature is far more rare, and far more dangerous than fever. Control of the cell sufficient to liberate enough heat or to permit of cell-nutrition (different for every species of animal), has evidently been reached only by great difficulty, foresight, labor, and delicacy of nutritional adjustment. Too much, within certain limits, can be guarded against, excluded, or compensated for, but a very little too little, is quick death. Empiric therapeutics knows that dangerous fever calls for cold, or other lessening of cell-metabolism. A quaint old physician ordered as his epitaph, " He fed fevers." That was one method, and of use, but many others are required, and our knowledge of cell-life must be much increased, before Biologos, when struggling with disease, is not as much hindered as He is helped by our clumsy therapeutics. The same lesson comes out clearly in the fact that the lower functions (muscular or glandular,) may be deprived of their blood-supply for considerable periods without injury. Their cell-compounding is evidently not so complex, and so much does not depend upon it. But deprive the cerebral consciousness-centers of their nourishment for the fractional part of a second and recuperation is impossible. The ceaseless attention of the engineer, his straining and sleepless watchfulness, the uninterrupted stream of God's life that must be poured into the mechanism of the nerve-cells and centers, (for He is steam and engineer in one) is beyond the limits of any finite imagination. Such is the price He pays for control of material cell-systems, such the reluctant consent to serve on the part of matter, built up into enormously complex and artificial tiny solar systems, and ever silently seeking to slip away from the Divine Master. Can it be believed that the intelligence and ingenuity that could do all this has no object, has not enough sense to have an object, in doing it? Blind evolution? Pah!

In every class of animals and plants the problem and labor of temperature-preservation is a different one. Each group has its peculiar range, above and below which control of cell-metabolism is lost. In a general way the lower or more simple the function, the simpler the cell-mechanism and the wider the range of temperature-change, the lower the average degree. Plants have been trained to periodicity of function, crowding the active work of life into a few of the summer months, shedding their lungs, etc., and, as it were, during winter, stilling down all tissue-metamorphosis to the lowest degree, they preserve the mother-cells intact, ready to redevelop all the destroyed organs when outside temperature permits, and with Spring's coming to reinstate the order and machinery of their inner purpose. The hibernating animals in their way have a similar but far more limited power. The entire machinery of natural clothing, bark, hair, feathers, and multiple modifications of skin structures, developed in response to the need of temperature-uniformity, and of withstanding the increased radiation in cold weather, has been not only an enormous tax upon energy and ingenuity, but has, of course, been a great factor in governing the character and direction of all biologic evolution. The few degrees of permissible temperature-change in the higher animals and in man, shows the complexity and delicacy of equilibrium of the typical cell, and especially of the higher nerve-cells, and indicates the heightened task, together with the necessity of compensating for climatic changes and thermometric differences by artificial clothing, housing, and fuel, in which the work of humanization and civilization so largely consists. Just here we catch suggestions of how the laws of cell-activity and cell-function dependent on temperature, reach out into all the institutions and developments of sociology and commerce, dictating very decidedly the character and work of government and the evolution of civilization itself. If God could control and preserve cell-function more easily, ours would be a very different world. All work and outlook is rigidly conditioned on the enormously difficult task of keeping His little instruments in a delicately poised temper and temperature of flux and obedient plasticity.

That Biologos works only through the cell-mechanism, and that through this mechanism he does work intelligently, is exquisitely illustrated by the healing-processes of wounds. A healing and knitting wound is quite as good a proof of God as a sensible mind would desire. Every tissue of the body, even the bones, has this beautiful power, and in every instance there is an adjustment of means to ends, an intelligent adaptation to the ever-varying circumstances, that is consoling to behold. A great surgeon has said that in some cases where callus would be useless, a subtle wisdom seems to recognize the fact in advance, and none is thrown out; but that when it would be of possible value, heroic efforts are made to heal the fracture, and with superabundant callus. Of the general means of healing, however, there is not the perfect disposal that existed in the nascent or young tissue. In scurvy and in other diseased conditions of the general system, old healed wounds will open again, and scar-tissue is evidently "making the best of a bad job." In plants there is the same power of healing injuries as there is in animals, and the similarity of the two processes is very striking. With a nervous mechanism or tool to help, the two processes would doubtless be identical. Much study has been given to the healing of wounds in animal tissues, but little or none in plants, which, interesting enough for its own sake, might also be of use in throwing a side-light on human surgery. A surgeon-philosopher of vegetable life is needed.

The entire result of bacteriological study and "Listerism" is summed up in the very simple advice: "Keep the healing wound clean and free from any interference of living or inert material that will hinder the healing process." The wisdom of the tissues will do the healing. The surgeon's duty is to keep the interferences away, and to help nature mechanically. Except cleanliness, all that can be done outside of mechanics, or the moving and removal of masses, Biologos will do; because, except indirectly through muscular contraction, He is incapable of effecting molar motion. The action of the concerned and related structures, or in the case of traumatism or surgical disease of the entire individual, is often and almost always indicative of powerlessness and subtle ingenuity in trying to circumvent it. That is to say, God's agent, man, endowed with intelligent disposal of mechanical energy, denied to God directly, is silently but longingly called and waited for. Man's function in the spiritual and biological evolution of cosmic purpose is to aid and supplement His work, to do as helper, and mechanically, what Biologos can not directly do. Medicine and ethics alike rest upon that basis. Thus co-working, these two, Biologos and the surgeon, have established modern surgery, and the splendid bravery of the surgeon has always been followed up, completed, and complemented by the divine loyalty of the inherent healing and curative power resident in the tissues themselves. In the history of every surgical operation, there shine between the lines and behind the facts, the divine eyes of intelligent love, denied mechanical power, and calling for the hand of the loyal surgeon for a moment's "lift." I had gathered but may not describe a number of striking instances illustrative of this dumb appeal of God, and failing to get it, of His heroic roundabout methods, the best with the means at disposal, taken to reach the end that intelligent control of mechanic forces could give in a minute. The extrusion of a renal calculus through the wall of the renal pelvis, and on through intestinal wall, thence naturally got rid of, with healing of the two vitally dangerous wounds; the cutting by peristalsis of several feet of intussuscepted bowel, and its passage per rectum, with reinstitution of the intestinal lumen; the encysting even in the brain itself of foreign bodies that could not be pushed out, and thus rendering them harmless; the history of every sequestrum; the long foresight and precaution against possible accident shown in the allowance of means for collateral circulation;—these and a hundred more—indeed, as I have said, every surgical case, to the sympathetic seeing mind, speak the lesson emphasized. Many volumes have been written and lives spent to discover the philosophy of inflammation, but the essence of all is that the intelligence of the cells rushes with loyal response, like watchmen and repairers to a dike-break, to repair the injury with all the zeal and means at disposal. Inflammation has been called "a physiological process strongly exaggerated," and Germain Sée has defined it as "a struggle for life, not a destructive process, but essentially a vital phenomenon, eminently reactionary, against a morbid agent." The function of the white blood-corpuscles, and that also of phagocytosis, show the same sanitary intelligence of the cell.

The key of almost all the tormenting mysteries of our

life is, I have said, the recognition of Biologos, *i. e.*, combined mentality and life, incarnating Himself in material form, by the sole mechanism of the cell, and contending with the infinite difficulty and labor consequent upon poor, obstinate, and dead material, and upon untoward circumstance. Try that key in whatever lock closes you out from intellectual, ethic, esthetic, biologic, or sociologic satisfaction and answering, and you will be charmed to see how the cosmos glows with light and comprehensibility, when before we stood in dumb suffering against the dead walls of fate, non-understandableness, and despair. If my own twenty years of heart-hunger, mind-hunger, and stoical abiding in the inevitable, have taught me the value of loyalty to truth as seen, the end of the imprisonment has also taught me that modern thought is needlessly, blindly, and illogically atheistic and pessimistic, or uselessly, formalistically, and irreligiously theologic, and that by the aid of natural sane human perception, this key will unlock for us an understanding of the living purposive world, itself the very flesh and body of a living, loving Father.

ORGANOFACTION IS THE MECHANICALIZATION AND AUTOMATONIZATION OF CELL-FUNCTION.—The setting apart of a number of physiologic units to perform one task is evidently primarily a simple method of conjoining the separate labors of individual cells, and thus massing their work to a common use and end. Their specialization, based upon the possession of inherent capacity by the undifferentiated cell to perform in a limited way any possible function, is a further striking extension and proof of that economy of labor that has been remarked by every biologist. The differentiation of cells into contractile elements constitutes the muscular system, and thus Biologos indirectly reaches mechanical power, otherwise impossible to Him. The independent control of mechanical force endows man with responsibility and power, but its derived nature necessitates the conclusion that man must hold its use in trust for and according to the purposes of its giver. In the past and present transition-stages of development, Biologos commands the service by instinct, poverty, religion, custom—all such being devices that enable Him to keep some control of the deputed power, until by intellectual perception combined with loyalty, man shall *choose* the right use, when the gift will be made more unconditional, and man will then become the intelligent and free co-worker with the divine.

But there is one aspect of organofaction that to my knowledge has not been recognized. I refer to the evident fact that it mediates an automatonization, an ordering, or an habitualization of function, with retainment of superior control by the great Mechanician, exactly analogous to every use of the body by individual man. Every habitual act or art of man is first performed with conscious volition. Slowly the first difficult steps pass into unconscious habit, the mind always occupied by higher, and as yet unlearned stages, but retaining a slight consciousness of the lower automatic actions, and always capable of instantly seizing them and directing them with conscious purpose. Thus the mind teaches the mechanism, as it were, to run itself, or delegates to subordinate nerve-centres the performance of repetitive functions whilst devoting attention to higher objects. I think there are many facts in life and biological science that go to show a precisely similar method on the part of Biologos. The most striking illustrations that occur to me are teratological facts, where an accidently-missed step of the process, a "failure of connection," as it were, or some other yet unknown cause, mars or complicates the higher design, and the automatic machinery goes on to the completion of a monstrosity in which a "slip," or a glaring loss of control of the subordinate working-mechanism is palpably evident. In certain tumor-formations a similar lesson seems evident. A bunch of cells "told off" for a specific function, gets misplaced, the organogenetic function loses control of them, seems thereafter to forget them, and to become incapable of hindering them from their prearranged development, and they run their automatic course of life. Pathology is proverbially faulty or morbid physiology, and pathology has many such negative or positive illustrations of this biologic device. The persistence of unused structures long after their possible function has been forgotten and undiscoverable, will to many minds suggest this as the explanation.

It should not be forgotten, however, that the possible vicissitudes and catastrophes of the future may make Biologos slow to utterly do away with old mechanisms that may again be required. So doubtful and apparently absurd as it may seem, we may yet be compelled to return to the primitive "natural clothing" of the animal, and if so the partially atrophied hair-bulbs would be at least a happy economy.

A positive example of the truth I am seeking to bring out is a well-established " law " that the longer an organism has been subject to uniformity of condition and action the less capable is it of adaptation to changed circumstances. Habit, of course, is a habit of the component cells, and the uniformity of cell-action renders them less plastic to the control of their extra-cellular master. Too long or too great uniformity of condition is therefore harmful to responsive plasticity and adaptability. But precisely on the automatonization (differentiation or habit) of function must rest the certainty and uniformity of nutrition and of progress, planting ever higher functions and uses on the lower, whilst reserving control of results, and without return to conscious readaptations and changes of the lower. The automatonization must be progressive, if the perfection of personality is to be progressive. Now, incarnation in a world subject to the tremendous differences of condition of ours, every condition being of unforeseeably undue length, shortness, or severity, the obstacles to an orderly and progressive, or a desirably speedy perfection of the objects of incarnation become tremendous. History, geologic, anthropologic, or written, is strewn with the sad evidences of the many desperate attempts and failures. Only by such a way of looking, a way that is moreover justified by the facts about us, can we account for the history of biological and sociological development, and keep intact our faith in and reverence for an overruling and in-living divinity. Especially in human history do we find the most marked illustrations of how habit and custom, good and necessary as they may be to give bases for progressive advance, become by mere momentum and excess of automatonization the veriest engines of evil. That obedience to constituted governmental authority was once a need and duty of superlative benefit to the French people, there can be no more awful proof than the continuance of that obedience when peasant-hunting was an amusement of "nobles," and when that authority levied 82 per cent. taxes to indulge in Versaillesism. But repetition stares us in the face all the time, and our own modern method of concocting thousands of millionaires on the infamous plea of "protection to American labor" all the while inviting the outside world of "pauper labor" to come here and compete for lowered wages and buy manufactured articles at advanced prices,—a happy illustration of the machine running away with engineer,—is also an unhappy illustration of the possible depth of human diabolic selfishness. I do not argue either for or against "the restriction of immigration"; I argue only against the hypocritic illogicality.

In the last analysis this phenomena of the automatonization of function is a mere consequence of the specialization of cell-function. The momentum of social forces carrying the agent past the point of objective utility, constantly seen in historic events, has its origin in specialized cells outnumbering and outlasting the lessened need of function. Life never voluntarily releases possession of a foothold once gained, and the differentiated cell-agents partially denied either nourishment or proper activity persist in function after the need has passed away. Thus arises disease of the physiological as well as of the social body. It is one of the consequences of unpliable material and of fickle circumstance, that Biologos has not yet entirely conquered. The conscious loyal helpfulness of man when it shall awaken and respond, will wondrously aid in the restoration of the balance of "supply and demand," not only in the human world but in the vegetable and animal as well. "Teach me Thy way, O Lord," is the prayer of the good man, whether he be meek or mighty. The bad man says there is neither Lord, nor the way of the Lord, and derides the most beautiful characteristic of teachableness.

The theory of De Vries, the Dutch botanist, also lends

support to the view of a diffused intelligence using all parts and cells of the nascent organism for metaphysical purposes and by metaphysical means. The latest researches in cellular physiology and histology are slowly putting to one side the theoretical, cumbrous mechanism of heredity devised by Weissmann, and the still more materialistic one of Darwin. Darwin's pangenesis-theory supposed that each of the different cells of the body gives off germ-cells capable of reproducing themselves, and that these gemmules penetrating the specific generative cells thus reproduce in the offspring all the peculiarities of the parent body. Weissmann supposes two classes of cells, the germ-cells and the body-cells, the first transmitting in unbroken continuity from generation to generation the hereditary characters. This ancestral plasma is therefore possessed of "immortality." The second class, the body-cells, composing the body-plasma, are used in building the tissues of the individual body. De Vries proposes to designate the cell in its entirety protoplast, and its different organs or parts he terms pangenes. All parts of the cell—not the nucleus or the chromatin rods alone but the chromatophores, chlorophyll-grains, starch-spots, membranes, walls, and even the vacuoles—all parts are living, vital, separate, independent organisms, of which the cell is the compound organism or colony. If this is true, as it appears to be admitted, there is plainly seen the invisible hand of Biologos at work differentiating, shaping, modifying, and using any and all means and parts of this little cell-universe. According to De Vries it is not the nucleus alone, but all the pangenes help to effect the hereditary transmission of organization. In offspring the pangenes grow into separate organisms of a character like those of the parent. What is this but to say that Biologos continues His work from the point attained, from the advantage gained, and with the cell-mechanism acquired. We proceed to a distant point by starting from where we stand. The cell as it exists must form the basis of the cell that is to be. Absolute and sudden change there can never be, only modification of the existing mechanism. Slowly but surely there is shaping itself in the minds of the genuinely scientific the necessary recognition of the truth that a metaphysical life is working out through the physical cell an extra-cellular purpose or intelligent will, which is cramped and hindered of free effectualization by the fact of cell-habit, cell-mechanism, and cell-nutrition. The frank acknowledgment of this extra-cellular mechanician would speedily resolve a multitude of mysteries and difficulties that now block progress, because of the impossible attempt to explain cellular physiology by purely material or physical principles. In many organisms the so-called germ-plasma, or, in other words, God's ability to reproduce an organ or an organism, is plainly scattered throughout the body. Precisely! and because God Himself is scattered throughout the body. In these less differentiated organisms the cells are still so plastic that working from where He stands, and with the acquired material, He can reproduce lost parts. Thus in the newt or lobster the amputated tail or claw will regrow, the snail's lost eye re-form, and an Italian naturalist, Mingazzini, has seen the entire brain of Tunicata, with all its processes, develop after its entire destruction. Every lover of flowers knows how a whole plant of certain kinds will grow from a small part, or a "slip" stuck in the ground.

CHAPTER V.

SENSATION.

THE mystery and wonderfulness of sensation have hardly been appreciated, often not recognized by physiologists. From the ordinary text-books one would gather that when a series of waves of air or of ether appears to us as tone or as color, it is quite as much a matter of course as when one pan of a balance goes up and the other down. But a moment's reflection shows us no discoverable relation or likeness between the stimulus and the sensation. There is as great or a greater difference between any two color-sensations as between the sensation of color and that of tone, and there is no hint to our minds why either should be what it is, or why a certain stimulus should produce a certain result. But to an alert mind nothing so inevitably forces the suggestion of a hyperphysical, designing, choosing, utilizing, ingenious artist and mechanician, as does the study of sensation. A materialistic or an atheistic physiologist is certainly the most illogical and absurd being that can be imagined, and yet one as much expects to hear from every second physiologist the sardonic sneer at "vital force," as to hear the heart called a pump or the lungs an oxygenating machine.

Every sensation and sensing mechanism shows a choice from an almost infinitely extended series of wave-stimuli of a limited, useful, and utilizable portion, and a rejection of those not useful or utilizable. Every one shows a purposive and a specific function of value to the organism, and such adaptations and modifications as are peculiarly useful to it. Every utilized stimulus appears on the mental side, or to consciousness, as something so utterly different from its original nature that, were it not for the sequence of stimulus and sensation, the latter would never be suspected of having any relation to the first. The mechanism intermediating the known cause and the mysterious effect is one that shows peripheral receiving, or end-organs, at every point of the skin, and from every particle of tissue, that lead to a unifying, responding, central intelligence, upon whose control and government the life and function of each part and of the whole depends. The mechanism is clumsily, but so far as it goes, accurately analogized by that of the switch-board and the city telegraph-office, which, from the local offices in every part of the land, receive accurate reports in cipher-messages, understood by no operator or clerk, of every external condition, and through this central governing office, again by cipher, orders and controls every distant part in accordance with need, desire, and circumstance. But as the writing and written telegram is unlike the transmitting instrument and Morse alphabet, as the electricity is an alien utilized force, as the manufacture of dots, dashes, and spaces first into the known-unknown jumble of Roman letters, and then by the solitary intelligence into meaningful and understandable messages,—just so is the crude stimulus of sensation passed through non-understood biological mechanisms, by means of foreign forces, and is translated into significance and usefulness only by the ganglionic centers and their manipulating mental governor. The importance of the nerves (telegraph wires), and of the central ganglionic or nervous centers (home office) is shown by the fact that embryologically these structures are among the first organs created, and that in death by starvation they are not touched. Neurologic life outlasts somatic or molecular life.

There are multitudes of facts in physiology and in pathology that show the highest biological control in closer and more immediate contact with nerve, ganglionic, and cerebral structures, and through them governing the semi-automatonized peripheral organs. The higher nervous structures are immensely complex in atomic constitution, and the greater the molecular complexity the greater is the contained force, the more unstable, and the more subject are they to delicate disturbance—*i. e.*, the more absorbed attention and unerring accuracy of adjustment are required of Biologos. Hence in death by starvation we understand why there is a preservation of the immediate organs of control and a feeding of these upon the muscles and less important peripheral organs.

Following the invitation a step further one may truthfully say that the whole body, the generative organs excepted, is but the mechanism of nutrition of the higher sensations and the brain. The body exists to furnish the cerebral centers with prepared food, just as the vegetable world viewed biologically exists to furnish the animal world with similar food. The higher is the last formed, the most difficult, and the most complex,—but it is just this that is most precious, and significant,—all of which, in biological development, shows the unrolling purpose, and the successive and progressive preliminariness of all the preceding stages. It is the last that alone explains all that went before, and it is the coming that will alone explain the present. God before all, through, foreseeing, and still preparing all, is profoundly evident. But it should not fail of observance that the purposiveness and intermediateness of every stage does not, as in the telegraph or any other analogy, reduce any structure to pure mechanicalism or to perfect automatonization. This is evidenced in the healing of wounds and numerous other facts. The subordination and differentiation of function are never absolute and complete, but at any instant, like any good military commander, Biologos may step down from the highest command and generalship, to lead, encourage, and use single cell-companies, or the smallest of " squads," with perfect command and with definite purpose. The consciously-acquired function never passes entirely into the unconscious or automatic. Representation and deputization never become forgetful of responsibility, because final authority is lodged only in one mind, whatever agents and helps are devised and allowed.

Such a chief help and agent is the nervous system, but, to the last correlating centre, its every particle is plainly though not absolutely an instrument, a tool, though not solely a tool, of the mind. Not solely, I say, because it is the unique quality of God, to overflow all He does with such largeness of love and plenitude of life that each incarnation, however subordinate, has joy in being. I am quite sure the tree and every vegetable growing thing has satisfactions, aspirations, delights, and a degree of consciousness that reward it for its limited powers and lowly uses, and is evidence of the beautiful divinity always larger than His work and ever having pleasure in His work. So in a higher degree with the animal, and so of course superbly with God's highest deputy, man. But even each cell and each organ has its own special and indisputable part in the silent song of all right-doing things. Whatever God touches is ever afterwards ennobled, and the veriest rag or piece of leather of the ash-bin is possessed of divine memories and qualities that arouse in the clairvoyant mind a thousand thoughts and suggestions of its far-way origin and history. Organization is like the photographer's sensitive plate of which, when once God has flashed into it His life, the dead particles will ever afterward show the lineaments of the divine face, until elemental decomposition is utterly complete.

A noteworthy fact about sensation is that with two exceptions its stimuli, of whatever kind, are vibrations. The first exception is the muscular sense, that of weight, tension, or muscular effort, the least useful and the least used of all the senses, so inaccurate and dumb that we see how difficult it is for spirit to reach any sort of relationship with masses and molar motions, even of the body itself.

In touch, the pressure sense, the stimulus is of mass, but of such slight degrees that it approaches the delicacy of the infinitesimal forces of the other senses, and, moreover, the molar force is undoubtedly at once received and mediated by the crowding of the molecular vibrations of the end-organs upon the delicately poised molecular balance of the true tactile element. With any high degree of pressure, there is speedy paralysis, and pressure of whole organs is perceived as the pain of injury, or of interference with organic function. All pain is the evidence of heightened or interfered-with molecular or nutritional function—always, one recognizes, a question of infinitesimal vibrations perceived or utilized.

But however caused, all peripheral stimuli must at once be transformed into nerve-force, of which we know little except that it is vibratory. At the ultimate center the agency of the transfer to spirit, the final mechanism of the relations of Biologos and matter, is beyond question that of the molecular and atomic vibrations within the highly organized cell, which is the means of influence upon or from the metaphysical.

We now know that the skin has limited areas and related end-organs, of course with intermediating and back-running nerve-fibers, for the sensations of cold, and yet others for those of heat. Impinging atomic and molecular vibrations below the health-norm in number are perceived as "cold," and those above the standard are perceived as "heat." But out of the infinite range of such degrees of vibration, only a tiny part or scale is chosen and utilized by the organism. These crude senses are plainly those of the lowest uses, such as defending, guarding, helping, and they have little psychic significance or value, except, and then in a still partial degree, when the mind is deprived of light and sound. In such cases there is possible a wonderful education and psychic extension of them. God can always as it were, re-enter and live again in his old dwellings, set them to semi-automatic tasks, remould and re-use his humblest still-plastic mechanisms, and make them shine with unwonted light and glory.

I once experienced an odd series of feelings that brought vividly to mind the fact that life, emotion, and happiness are compatible with great differences in and deprivations of ordinary mental stimuli. A social gathering was going on in a large hall, and the sounds of excellent music, dancing, laughter, and gayety came from this room where a hundred or more happy folk were passing happy hours. Upon opening the door and being ushered into the hall the room was found to be as dark as midnight. Not a thing could be seen. It was the social hour of the inmates of an institution for the blind! The first uncanny, creepy feeling was soon dissipated by the thought of congratulation that indomitable mind and spontaneous emotion, though deprived of light and vision, could still find satisfaction and play through the medium of indirect sensation. Soul still conquered sense!

The same thought is exemplified and emphasized by a consideration of the report of a convention of deaf-mutes. One is apt thoughtlessly to pass over the beginning of the report that says, "The meeting was called to order by the president, who rapped vigorously on the desk to attract the attention of his audience." An audience of deaf-mutes called to order by a noise! Those who see a blind man tapping the street in front of him, as he walks, are likely to think that this is solely to avoid objects that the cane may strike. It is also to avoid objects that the cane does not strike—because to the blind man's ears and hand there is a timbre from the blows upon the pavement near its edge, near posts or steps, that is very different from the resonance when the blow is not near such objects. It is said that blinded bats are able to fly unharmed, avoiding objects in their flight by means of the perception of an increase of barometric pressure of the air close to those objects—so sensitive to variations of pressure is the expanded interdigital membrane.*

A suggestion is indirectly aroused by this fact as to the "relaying," if one may so speak, of crude and faint stimulation by the mechanism of the nerve-ganglia and centers. There is a nervous device that reinforces and transforms whilst also repeating the subtle, weak, and in themselves

* The experimental blinding of the bat, however, was not necessary, because millions of bats winter in Mammoth Cave, miles from the faintest ray of light.

meaningless, hints of the external world that we call sense-impression. It is the living prototype of the electrician's "relay" and microphone combined. Thus all man's mechanic devices are but poor imitations and repetitions of what Life's vital forces have long ago brought to wondrous perfection!

In the deaf-mutes' convention prayer was said, the roll called, addresses made, business conducted, and long sessions held—all in the sign-language—all in silence! "The amended constitution and by-laws were adopted *after a lively debate.*" If present, our blind friends would certainly have thought the meeting very strange and stupid. But the success in raising funds for a proposed home for aged and infirm mutes, and the discussion of other worthy objects, made the gathering a very interesting one for the attending delegates.

According to the Paris correspondent of the London *Times* the method of analyzing motion by the chronophotograph, which has been so happily applied by M. Marey in the case of moving animals, such as horses running or birds and insects in flight, has recently been employed by M. G. Demeny, a preparator at the physiological station of M. Marey, to examine the movements of the lips in speaking. He has obtained results which show that the form of the mouth is quite different for the different articulate sounds. With these photographs combined in a zoetrope he has reproduced the movements of the lips by synthesis. An ordinary person finds it difficult to read the words by the animated pictures; but a deaf-mute who has been accustomed to read from the lips of a speaker, found it easy to do so from the photographs. A young pupil of the National Institute of Deaf-Mutes in France could read the vowels and diphthongs as well as the labials. The first experiments were of course, not all that could be desired; but, in bringing the matter before the French Academy of Sciences, M. Demeny expressed the hope that in continuing his researches he would be able to develop a new method of educating deaf-mutes by sight from more perfect photographic images. Obviously a magic-lantern lecture might be delivered to an audience of deaf-mutes in this way.

The encouraging and deeply suggestive fact of rescuing the faculty and power of speech in these deaf-mutes is one that commands our sympathy. There is no limit to the ingenuity of Life and to her triumphs over adverse circumstances and deprived stimuli. We have all read of another striking example—very different in kind, of course, but illustrating the same great truth. One of England's greatest statesmen was blind; so was a great numismatologist; and another of her great men, a hunter and rider of unexampled daring, a peerless sportsman, an excellent business man, and an active administrator, had neither arms, hands, legs, nor feet. One is reminded of the philosopher's answer to the Millerite who excitedly told him that the world was to come to an end that day: " Oh, well! " was the answer, " we can get on very well without it."

It would seem that if loss of sight were added to loss of hearing and speech, naught but tragedy and melancholy could be left, or that merely the routine life of the lowest functions such as those of nutrition, would persist. But there are few happier and brighter-minded people than Laura Bridgman was and than Helen Keller is. Another, a man likewise deprived of these great avenues of influence from and communication with the external world, without which life to us would seem so barren, travelled all over the United States alone, raised a family, and lived out his period of brave and satisfied life. He could talk with anybody by means of the ingenious device of tattooing the English alphabet upon his hand. Words and sentences were spelled out and recognized by the positions of the letters touched.

The emotional life of these imprisoned souls, cut off from so many relations and avenues of interchange with the external world, must be all the more vivid and hyper-sensitive. A coarse jar of the hyperesthetic receiving end-organ of sense is transformed into a rude thunder by the highly attuned and delicately responsive microphone of the inner sensation-making mechanism. Thus the possibility of causing sharp sorrow is a necessary concomitant of the ease of eliciting joy. It is the glory of civilization to care for such and shield them from pain, and it is the delight of medicine to minister to them its healing. It is hard to sympathetically understand and realize the inner life of these almost windowless minds. How strange must seem to them the dreams and somnambulisms of never-to-be awakened emotions, the dumb reaching-out toward reality of denied possibilities, the unsatisfied hungerings of imprisoned sensibilities. Their minds must be thrilled by dim hereditary echoes and the far-away caresses of ghostly ancestral hands. With what pathetic half-responsiveness do these shut-in souls catch the shimmer of long-departed life, that comes to them like the last faint evening flushings reflected from distant mountain-tops to valley dwellers that are in the night.

The sense of taste is a response to the molecular vibrations of dissolvable bodies, and its purpose and use are evident. But even with as lowly and temporary a function as this there is proof of self-satisfaction, a beneficence that surcharges pure use with a luxury of pleasure and of diversity, and is never content with leaving brute utility without a touch of useless but delightful heaven. If Biologos were satisfied with telling the eater simply what was good and what was bad food, varying degrees of one kind of sensation would have been sufficient warning. Bitterness, pungency, acidity, and a hundred other unpleasant tastes, would have been alike or with varying degrees of disagreeableness, whilst of pleasant things a thousand delightful-tasting fruits and edible goods would have been alike, or all simply savory more or less,—not, as now, with the exquisite differences as from utterly different orders or types of stimuli, that give a passing pleasure and invite to an educational examination, and to the creation of new forms.

Similarly, the perception of the vibrations of volatile bodies by the mechanism of smell is a function not only of utility, but has exquisite extensions, psychic pleasures, and uses, that recall the same kind elaboration and bountifulness. But our atrophied organ is indifferent to a million of existing perfumes. The uses of the sense of smell in wild animals are numerous and of an exquisite perfection. The odors of every animal, peculiar to each family and even to each individual, are the determining agents in warning against enemies, in preserving the hives of bees from intruders, in telling of the existence of water, or of better pasturage, as many, it is said, as five hundred miles away, in telling of approaching storms, in protecting (by disgust, etc.,) from enemies or noxious agents, or in avoiding toxic substances or malnutritional food. It is said that 1-46,000,000 of a milligram of mercaptan can be perceived by the sense of smell.

But in the uses and out-building of the sensation of

hearing we stand amazed at the exuberant and spontaneous life beyond any possible use. Gratuity is heaped upon abundance, and, with swelling heart, the mind through music, as Beethoven said of himself,* comes into close union with God. The perfection of the sense of hearing, and the relatively complete utilization of all lengths of aërial waves, provokes comment and inquiry. There is considerable evidence and much proof that insects and other living organisms have sensory mechanisms that react to aërial wave-lengths far smaller than our own ears respond to. The upper limit of perception of the human ear varies in different individuals, but below some 5,000 vibrations per second, our ears and auditory centers have a marvellous perfection of response to wave rapidity, wave-dimension, and wave-shape, respectively called by us pitch, loudness, and timbre. When one ponders upon the individual perception and easy analysis of the multitudinous synchronous tones and qualities of tones of a full orchestra, one is lost in wonder that the auditory mechanism can intermediate and unravel the whole.

And gratitude outruns wonder when we think of the magnificent unnecessariness of the gift. For simple utility's sake the whole could undoubtedly have been made infinitely simpler, danger foretold, and needed help gained as to the environment without the magical richness and endowment superadded to the necessary. This perfection of mechanism has been possible because of the combined strength and fluidity of the stimulus. Of the two gaseous oceans, the air and the ether, in which we live, the latter has acted with such incomprehensible small-

* " I well know that God is nearer to me in my art than to others. I communicate with Him without fear; evermore have I acknowledged Him and understood Him."

ness and lightness of force that it has presented the frightful obstacles to the elaboration of its reacting mechanism to which I shall allude by and by. But in the aërial waves there is a sufficient strength of force, and at the same time a sufficiently high degree of gaseousness or fluidity, to assure pliancy and rapidity of act consonant both with mental speed of movement and suppleness, and with the concentrated play of external event. The mind itself would most certainly be more at home with a stimulus and mechanism infinitely more subtle and rapidly-acting than air and auditory apparatus, but the difficulties of creating a mechanism respondent to such slight and rapid undulations is enormous and is exemplified in the case of the eye. To sound-waves, however, though for the highest mental use greatly too crude and slow, Biologos, with characteristic ingenuity, has a recipient and intermediating mechanism that gives us two wondrous languages of emotional and psychic life. Speech and Music!—what suggestions the simple words call up! How intolerably pathetic would our life be without them—apparently, and now that we know them—because,—Hush! —there ever recurs the inexhaustible idea of the love and richness of God, as the warning thought quickly breaks into our thankfulness that neither tree nor animal has these awful blessings, and yet are they happy, still are they ever satisfied. The compensations and bounteousness of blessing of the lower are not lessened by the superlative lavishness of the unexpected gift of the higher. So kind, so kind, is the dear Father of us all!

It is often said that music is the one creation of absolute novelty of the human being, and that herein man shows himself, as it were, superior to God. The thought would be unworthy of mention were it not seriously and persistently repeated by the human egotist. How He must smile at such childish prattle! Even if it were true we might at least thank Him for the instruments, for the permission, and for the outfitting of neurological and mental mechanism. "He that planted the ear, shall He not hear? And He that formed the eye, shall He not see?"

But the egotist did not know the birds. Even the few poor, dwarfish, prosaic little birds that hang to the fringes of our temperate-zone civilization show him gladness and music at the heart of life. But it is only in tropical birds, they that have not been blasted by the devilish impiety of man's cruelty, that can sing as God would have them sing. I would call particular attention to Hudson's exquisite chapter on music and dancing in birds and animals, in *The Naturalist in La Plata*. Among other things he says:

"My experience is that mammals and birds with few exceptions—probably there are *no* exceptions—possess the habit of indulging frequently in more or less regular or set performances, with or without sound, or composed of sound exclusively, often of the most complex and beautiful forms. . . . Many songsters in widely different families of birds possess the habit of soaring and falling alternately while singing, and in some cases all the aërial postures and movements, the swift or slow descent, vertical, often with oscillations, or in a spiral, and sometimes with a succession of smooth, oblique lapses, seem to have an admirable correspondence with the changing and falling voice—melody and motion being united in a more intimate and beautiful way than in the most perfect and poetic forms of human dancing."

Note especially his discription of the song of the white-banded mocking-bird. Of the spur-winged crested screamer, a noble bird as large as a swan, he says:—

"Its favorite pastime is to soar upwards until it loses itself to sight in the blue ether, whence it pours forth its resounding choral notes, which reach the distant earth clarified, and with a rythmic swell and fall as of chiming bells. It also sings by night, 'counting the hours,' as the gauchos say, and when they have congregated together in tens of thousands, the mighty roar of their combined voices produces an astonishingly grand effect."

Hudson rightly judges that the songs and flights and dances of birds and animals are not, as Darwin would have us believe, due to erotic fury. He says they are due to the genuine pure gladness in the heart of the bird. For example, the male birds often arrive first in their migrations, and even with the female absent and pairing time a month hence, their songs burst forth most divinely. Wallace also, together with Ruskin, agrees that brilliancy of color goes with vigor of life and purity of substance.

But the egotist also forgot to say who created the mechanism of the human voice and of song, the memory of melody, and the grief or the gladness of heart that finds expression in the pathos and splendor of music. For purposes of speech alone, for the simple necessity of the mechanical intercommunication of ideas, a monotone is all that is necessary. But what a glorious superabundance of gift in the octave-ranging, palpitant richness, and soul-expressiveness of the human voice.*

* Probably as ludicrous a thing as ever happened was the experience of my friend, Professor Roswell Park, of Buffalo, N. Y. A man's life had been saved by the beautiful surgical skill of a successful laryngectomy. When health had been restored, my friend proposed to his patient the insertion of an artificial larynx, so that vowel-tones, or true voice could be added to the whisper that necessarily resulted from the absence of the vocal cords. This apparently highly desirable thing was done, but the tone, of course, was uniform; there was no change of pitch possible to the mechanical larynx, and expression, modulation, timbre,—everything that makes voice pleasant and

But over and above all such human conceits hovers the absorbing thought:—What man does is God's doing, because man is His work and is Himself at work. Whatever is the honor and glory of the one is that of the other, and man should be as willing to acknowledge the symphony and orchestra to be the labor of the Divine Artist, as that magnanimous musician is glad to allow the honor to His dear children, in whose music He not only rejoices, but in which He is theme, harmony, and player at once. "He is the Singer and the Song."

Between waves of a few thousand a second up to those of some four hundred millions of millions a second there is a blank space that is so huge in extent that it suggests the possibility of a million orders of sensation, each as different and each of as large range as either that of hearing or that of seeing. If only there had been the stimuli! If, instead of two oceans, we lived in a thousand, each of different degrees of fluidity, and if we were outfitted with a sense-mechanism for each as beautiful as that for air-waves and that for ether-waves! Eternity would then be too short to exhaust the pleasures of sensation, each type having its own world of opera and composition, as diverse and as progressive as that of music, or that of painting. In a world, the specific gravity of which is different from our own, the

more than useful was absent. The man could speak, convey ideas perfectly, but when he tried to give emphasis, *nuances*, shadings, diverse meanings, and especially when he tried to express emotion, anger, or resentment, there was only the monotonous drone and squeak of the intolerable machine. Nothing could control the convulsive laughter of surgeon and assistants. The poor man's indignation sought outlet in speech, but the very words of wrath were turned to outrageous absurdity by the infernal device. In a spasm of ebullient rage he tore the mechanism out of his throat, cursed the man who had saved his life, and is probably running and hoarsely whispering invectives at him still. He never came back.

atmosphere of a different density, and with other conditions of life varying from those in our own globe, the size, appendages, and conformation, of the body of the inhabitants would necessarily vary largely from our own; but it is in the differences of possible sensation that the greatest variations would occur. Should such, to us, new worlds of possible sensation exist out there, we would never understand them, because it is impossible to tell another what a sensation is. Indeed, in our sight and hearing, extension beyond the limits already set is hardly possible. In color-sensation how we desire eyes that would react to waves above and especially octaves below our present confined ability. There are the waves, steadily pouring upon our eyes, but wholly unutilized,—nay, the eye has even to shield itself from these rejected parts of the spectrum by special devices. The entire series of magnetic and electric waves constantly streaming through and about us are also unutilized, unless the suggestion I have thrown out be true that the Homing Instinct is carried out by means of a mechanism responsive to the magnetic currents of the earth. The reason for this neglect to devise a reacting sense-mechanism to these waves, is because of their changeability, the inutility of the function to the developing organism, and the inefficiency of the stimulus. This will be made more apparent later.

There are many facts that prove that the development of the function of vision has been extremely difficult. The function is not only dependent upon, but it consists in a retinal reaction to fine degrees of heat, the light-waves of the ether being translated by a sensation into color, according to their sizes or heat-energy.

The exclusion from vision of the ultra-violet and the infra-red rays of the spectrum is because of their relative weakness and inconstancy. As to the ultra-violet, the insignificance of the total energy of these rays is perceived by noticing the tiny curve and extent of activity of these undulations figured in the little triangle beyond H of a normal spectrum. Not only is the energy too inconsiderable, as compared with that of the adjacent red space, to produce sufficient organic reaction of the ocular mechanism, but the varying quantities of such rays, resulting from the ever-changing remnants left over after the absorptions of the atmosphere have been satisfied, render constancy of stimulus impossible, and the eye naturally fails to develop response to such stimuli, or it is forced to exclude them from the retina, because they are so weak and capricious. Moreover, we must not fail to recollect the exaggerated importance and mistaken supposition of extent of the rays of this part of the spectrum, naturally engendered by their influence in photography. A line must be drawn somewhere, and the remarkable fact seems to be that the violet rays should not also have been excluded from color-production. The limitation of the chromatic scale should not be a cause of critical complaint, but its extension is rather a source of delighted wonder.

The infra-red rays so rich in the total amount of energy, and so magnificent in the range of their wave-lengths, do not produce visual results because though three fifths of the total energy of the spectrum is in this region, it is spread over a space six times as great. Now it is apparent that amplitude of range beyond a certain necessary and satisfactory amount, would be of no advantage. Extension of the visual function to these regions would imply immense complications of the ocular and cerebral mechanism, complications and intricacies that, if in any way possible, would hardly give any considerable advantage to their possessor.

In the next place we must note that this great stretch is one continuous succession of peaks and valleys, or in other words, that its energies are not constant and regular. Cool places alternate with hot places, and though the shorter wave infra-red peaks are as high as some parts of the visible curve upon the other side, they are not regularly so, and during a fall of the general supply of the radiant energy, as in an obscure day, or in the morning or evening, these valleys would drop down to a level where their maximum of energy would be very small or even *nil*. And this fact of a continual pulsation of the general amount of the stimulus must be borne in mind. It is not sufficient to say that the great space from 0.80 to 0.90 contains an energy as high as that between 0.40 and 0.50. Upon this last side of the curve there is no valley at all, whilst just within A we have a tremendous one that with diminished general illumination would make a break, sharp and decisive, between the two visible portions. The existence of this deep and cañon-like chasm at A seems to me highly suggestive. At this point there is the boundary-line of visibility. Did this cool notch not exist, it is easily conceivable that the retina might have learned to react to the stimulus extending so far as wave-length 0.95, or even to the next great cool space at 1.15, had there likewise been no such fall as is seen at 0.95. But the great ebb and flow of radiant energy every twenty-four hours, complicated always, and rendered more highly inconstant by the varying conditions of atmospheric cloud and dust, of shade, of vegetation or hill, of longitude and equinox—all these sources of variation and inconstancy proved to the hard-tasked ocular and psychic Mechanician that He could rely with confidence only on that part of the spectrum, or upon those wave-lengths that under all conditions were strongest and most persistent. His creatures had to gain the best intimations possible of external objects during foggy and cloudy days, in twilight, moonlight, nay, even by starlight, and at these times the reduction or even the extinction of energy in the infra-red regions would be too great or too frequent to give such indications. Hence the necessity of ignoring them.

Professor Rowland has ruled 160,000 lines to the inch in his concave gratings. The finest slit admitting a beam of light to such a grating is one fifth of a millimeter wide, yet this covers more than the distance between the two D lines, and in this space there are at least a dozen alternations between brightness and extinction. The astounding sensitiveness of the bolometer is far from discerning the absolutely homogeneous ray. It has been shown (by Nichols, *Am. Jour. Sci.*, Oct., '84.) that the retina responds to an exposure to rays (from the pigments of a Maxwell-disk) lasting only 0.00144 of a second. This was the time for yellow, and at the brightest of three degrees of illumination. For red the figures were 0.00209, for violet, 0.00286. Experiments with spectra of the solar radiation are still more remarkable. B. O. Pierce, Jr., (*Am. Jour. Sci.*, '83) found that the largest displacement needed by any observer corresponded to a difference in wave-length of about 0.000005 *mm*, the smallest to a wave-length of about 0.0000005 *mm*. In all these cases, the extreme refinement of response is in yellow—precisely where the stimulus is strongest, most invariable, and most persistent. The most trained imagination can catch only the most impotent and inadequate fancies and glimpses of the marvellous sensitiveness of the retinal mechanism that can react to stimuli represented by the foregoing figures.

Sunlight, we are told, is composed of the following parts:

Red..	54
Orange-red.................................	140
Orange	80
Orange-yellow..............................	114
Yellow......................................	54
Greenish-yellow	206
Yellowish-green............................	121
Green and blue-green.......................	134
Cyan-blue	32
Cyan..	40
Ultramarine and blue-violet................	20
Violet	5
	1000

Condensing the intermediates with the principals we have

Red colors.................................	194
Golden colors..............................	454
Green colors...............................	255
Blue colors................................	97
	1000

If we ask what great color-classes of visible objects have most occupied man's eye and mind in all past history, the first in overwhelming importance is light and fire; the second, the world of vegetation; the third, blood, as the concrete representation of war, struggle, and superstitious symbol; the fourth, the sky above with its reflection in the waters of the earth. It would be difficult to name another class, for whatever other colors the world may have presented to the eye of historic man, they must have been mixtures of these, or unimportant exceptions that have left only a small and inconsiderable organic response in the psychic mechanism. The ordinary suffused daylight, even of a clear day, is slightly yellowish, and in almost all degrees of obscuration of the sun, the more refrangible rays being cut off, the indifference-point of the spectrum is sent farther down into the yellow band. Whether from a greater turbidity of the atmosphere, or from an increase in thickness of the several-mile-deep dust-shell always present over the earth, or from the morning and evening obliquity of the sun's rays, much of the *day* of average humanity has been more yellowish than was perhaps mistrusted. This, it is probable, was more pronouncedly the case in the earlier stages of the world's life. Moreover, the rising and the setting of the sun have always flooded the earth for one or two hours each day with a glory of orange or of golden radiance. It is also certain that our earthly fires are of a ruddy golden or yellowish hue, and the student of mythology and of early religions knows well enough the part that fire has played as a representative of the unseen divine life, or as an analogue of the recurrent changes of the lights of the sky by day or by night. "Pyrolatry," says a life-long student and historian of ethnic religions, "is common to all religions." "Through the whole history of Aryan faith runs the fire-symbolism of Mithra." Quotation from a thousand sources could be added, all of the same import. We all know the beautiful myth of Prometheus and the stolen fire. Ever since man's dawning intelligence caught a glimpse of the mystery of light, of the wonder of the strange lurid glow of the sun at eventide—nay, even back to the time when, by its aid, he cooked the flesh of the animal whose blood he had shed, the wonder of fire was daily and hourly before his eyes. Further reason for including the daylight, the sunshine, and the fire-hues under the general term golden, comes from the symbolism of gold itself, which, in all ancient faiths, as well as in the instinctive feeling of the modern artist and poet, is the representative metal and color of the divine glory and halo.

To this consensus of reasons might be added the comparative absence of whites in nature. Clouds are sometimes a dull or grayish-white, and snows, however considerable in some countries, are certainly the world over, a small and short-lived covering of the earth's surface. Wherever white sunlight falls on land or tree or rock, it is always reduced to colors by the unequal absorption and reflection following; these colored reflections are the eye's customary stimuli. When sunlight falls on the sea, only a small portion of the surface reflects white back to the few eyes there or thereabouts. So that as a fact white sunlight is generally reduced to yellowish tints, or to other shades before it reaches the eye. Where this is not the case, the rays would be too powerful, and unpleasant effects upon the eye would be certain.

The closeness of relationship between white light and golden light is also shown by the ease with which spectral yellow, by increase of illumination, passes over into white, being, as it is, the nearest of all colors to the luminous intensity of that compound. Consequently a complementary color of the lower kinetic value is all that is required to quickly heighten it into the white to which it is so closely allied.

The proportion of the spectral golden rays, 454, or nearly half of the whole, represents the overwhelming part the diffused lights of day and those of fire have played in the world's history. The unity of character running through this vast space of the spectrum, testifies to the unity of the cause, and to its power both physically and mentally.

The vegetable world—whose greens take up the next greatest portion of the spectral rays—representing one fourth of the whole—is so plainly the origin of the green band of the spectrum that it is unnecessary to go into detail concerning it. When eyes appeared, next after the golden light of day, they would certainly fall upon some of earth's verdure, and except to the city-man, the proportion holds up to the present time. Green is philologically the growing thing, and grass or tree covers the face of the earth.

Red occupies the next lower degree in the proportion of the spectral waves. The crimson of the fruit man ate, or of the wine he drank; the deeper orange hues of the flame-points or embers of his hearth-fire; the autumnal red of the forest trees, or the expansive glory of an occasional scarlet sunset, would not, all combined, account for the proportion of spectrum-space occupied, and these things are infinitely far from explaining the intense and distinctive character of the subjective sensation of spectral red. I believe that it can only be explained by the rôle that war and bloodshed, blood-sacrament and blood-rites, have acted in the history of the race from man's egress out of animalism and progress to nineteenth-century militarism. "The blood is the life." In a curious and deeply instructive book, *The Blood Covenant* (Trumbull), one learns something of the influence of the vision of blood-shedding in the early world. It is an instructive, though a ghastly picture, that, despite the author's sympathy and sanction, makes one shudder. Strange insights, these, into human nature, that we gain in reading of the blood-drinking, blood-bathing, blood-ransoming, blood-unions, blood-compacts and friendships, blood-sacrifices, and blood-suppers, blood-burials, blood-cures and sprinklings, bloody hands and uplifted arms, blood-transfusions, human sacrifices and cannibalisms, bloody passovers, and blood atonements. What an echo, too, of long-passed ages when bloodshed was no mimicry. The bloody idea is certainly " nail'd wi' scriptur'."

But this it may be said is all legend and myth. Yet mythogenesis is organogenesis : when beliefs were making, organic structures were hardening and shaping, and moreover when authentic history begins, it writes of the sword and red-handed death ; the record rolls on with the tired centuries, depicting one monotonous tale of sanguinary strife. " War is the matter that fills all history," says a great historian. One million nine hundred and forty-eight thousand lives lost in the last twenty-five years in European battles, is the last record, with Europe a huge camp to-day.

The proportion of spectral blue is small in extent and weak in power, and it has a character of distance and of impersonality exactly corresponding to the sources whence this color has reached the eye. The sky is above, but man's eyes are seldom raised to it. At the horizon it often fades to the violet in which the spectrum likewise passes out of sight.

Certain colors are called primary or elementary, because they have been derived from these great divisions of natural objects that have been reviewed. They have been the uninterrupted stimuli of the visual function since the brain sent its retinal servant out to the body's surface to see by its aid. Some, if we may so speak, are more " elementary " than others, in the sense that some stimuli have been either more prevalent, more powerful, or more interesting than others. This is overwhelmingly so of red and gold. In Swinburne's *Poems and Ballads*, Mr. Grant Allen found the red epithets numbered 159, the gold 143, the green 86, and the blue 25. In Tennyson's *Princess* the respective numbers were 20, 28, 5, 1; and it was so in other cases.

A "color" of the spectrum occupies just that amount of space, or, to put it in another way, waves of more or less extended differences of length are perceived as a single color, just as the bulk of the waves from each of these classes of objects have been most uniformly and persistently reflected into the eye during the growth of the race. Nature has acted upon the organism in these continuous ways, and the cerebral products are the spectral colors, in the proportions and with the characteristics we find appearing in consciousness. The largest and most persistent stimulus has been that of the gold rays—the varied shades of the diffused light of day, or the ever present mystery of fire. These have been poured in profusion into all eyes, comprising nearly one half of their total stimulus, while the green rays make up a fourth, the red less than a fourth, and the blue a still more limited amount.

It is a remarkable fact that the objective luminous power follows the same law, and is not caused, as we might *a priori* suppose, by the wave-length. According to the measurements of Macé and Nikati, the following are the relative luminous powers of the wave-systems corresponding to the wave-length, the highest power corresponding to wave-length 569 μ (yellow) taken as unity:

681	656	641	613	589	569	550	534	527	520	507
0.015	.080	.111	.252	.768	1.000	.954	.512	.400	.314	.128

In order to reach the same result of visual acuity, as in the 569 wave-length rays, the quantity of light of red had to be increased sixty-six times, of green three times, of blue eighteen times, of extreme violet 5,460 times.

The explanation of these figures will be found to lie in the physiology of the retina. The stronger or longer waves do not, as we see, produce the most powerful effect; indeed the luminous intensities have no relations with refrangibility, but seem to depend on facts of another order, or the utilization of the residue of rays left over after the absorptions of natural bodies have been satisfied. The greater the wave-length the more complete the absorption, until the line of the descending curve dips pretty low down in red or orange, when the residue becomes so great that the light-curve takes its swift rise, to fall gradually from its crest in $D \frac{1}{4} E$, to the extreme end of the visible spectrum, where the dispersive forces of atmospheric refraction allow few of the more refrangible rays to pass to the eye. So the retina has learned to react, not to the most powerful, or to the finest, but to the most continuous and steady stimuli. Its response, therefore, is more perfect to the gold rays, next to the green, lastly to the red, and to the blue, as Macé and Nikati have found.

But there has been a great failure to differentiate the objective from the subjective intensity. Confounding these wholly different phases has resulted in the *non sequitur* of Magnus and Gladstone, who think red was the first of the colors historically developed. Color being a creation of the mind, and after a double transmutation of forces, it follows that its subjective character may in part be independent of objective causes, or of direct stimulation,— may be a complex whose elements are by no means all gained through the retina or the visual mechanism. A thousand facts prove that of all the senses vision is the most free from bonds of logical and necessary connection with the primary sources of stimulation. No fact is more strikingly characteristic of this law than these differences between the objective and the subjective intensities of colors. As we have just seen, the order of the former is the highest in gold, falling then to green, to red, and finally to blue; the order of the subjective intensity is first red, then golden, then green, and lastly blue; and precisely this order tallies with that of the vital and personal connection with man's life of the four classes of natural objects we have named. The mordant acids of life's needs and passions have eaten these tones deeply or less deeply into man's brain, according as they have in varying degrees been associated with his miseries and gratifications. Only on this principle can the vivid and powerful effect of red be explained. Private and public bloodshed, social and religious blood covenants, all of which have always been bound up with every day of humanity's advance,—these have bitten into his being an intensity of response unapproached by all other chromatic stimuli. In the language of physiological psychology, this fact might, perhaps, be expressed as the demand for more numerous connections with other cortical centers, corresponding to the variety of interest the stimulus excites in them, and the power required in co-ordinating the multitudinous waves of emotion called forth.

The most valuable thing to a man is, of course, his life, symbolized forever, in fact and in covenant, in rite and in ceremony, by his blood. Next to this comes the light of day and of fire, which he has always represented to his mind, as it has been to his eye, of a golden hue, under which term may be accurately grouped the changing effects of the ruddy orange, or the yellowish whites of light and of fire.

Among earth's vegetation man has, of course, built his home ; but there is in the subjective green a lack of power and intensity exactly corresponding to the nature of our impersonal and semi-independent relations to the meek verdure and growing things about us. In blue these qualities are exaggerated into the feeling of distance, coldness, and elevation, derived, of course, from the faraway mystery of the sky and sea.

The intermediate colors of the spectrum should be considered for a moment. The fact of their existence is almost forgotten by color students. This neglect is all the more remarkable when we observe their amazing extent. While each pure " primary " color comprises from forty to eighty parts, we find the mixed intermediates stretching out to 140 between orange and red, to 114 between orange and yellow, and to 327 between yellow and green. Strictly speaking, these are just as "primary" as the other shades we call red, or yellow, or green. The whole nomenclature is relative, a mere thing of custom or arbitrariness. If simple spectral space occupied, or, if the proportions these intermediates bear to the whole number of rays, were decisive, the small spaces of the purer colors would serve as the unnamed delimiting lines for the other and larger stretches and quantities. The extent of these spaces shows us how differing nature's "colors" are from those of the mind, or rather, what receptacles and constructions the mind puts upon the color-intimations or wave-hints of nature. The "colors" of the inorganic world are always broken and mixed ; the spectrum gives us homogeneous wave-systems, sorted out of the compound, and arranged seriatim. The prism brings order out of chaos, whilst for itself the mind still further idealizes and reconstructs another world out of the spectrum, by ignoring the mixed intermediates, and emphasizing the small spaces more pleasing to it. But, keeping close to nature, we must ask concerning the significance of the extensiveness of these spaces, and this can only lie in the fact that external colors are not saturated (from homogeneous wave-systems of maximum strength) but are always from mixed wave-systems, culminating in a higher average of those of one of the four primary colors in each of the four classes of phenomena mentioned. The ocean-swell may be made up of many lesser crests and troughs, but there is always one point where the general variations reach their maximum, and this would correspond to the narrow limits of the *pure* color. But between these crests are large regions of indeterminate mixture. Between the pure hues of the deep autumn reds and the paler yellows, and beyond the rapid instants of ruddy flames and setting suns, are the multitudes of the ever changing tints of brighter glows, which account for the 140 parts between spectral red and orange. In a like manner we perceive the rationale of the 114 parts between orange and yellow, whilst the protean changes and mixtures of the ever varying light playing amongst the myriad-tinted shadows of the infinite variety of vegetable forms, produce the enormous interspace represented by the 327 intermediate parts between the yellow and the green of the spectrum.

It will be seen that the endeavor has been to institute a correspondence between luminous stimuli from the natural world, and the chromatic effects of the spectrum's analysis upon the mind. Our color-sense must be the organism's response and reaction under stimulus; in a word, it must be investigated by the methods of study that in all other departments of biology science has taught us to use with such brilliant results. The hand of a man, the wing of a bat, the dog's fore-foot, and the horse's fore-leg, the bird's wing, and the seal's paddle,—these are all modifications or variations of one fundamental structure, according to the work to be done, and in response to the peculiar stimulus; just so the cerebral products of multiform color-stimuli have left their psychical homologues in our own complex color-sense.

In accordance with this conception of the origins of our color-sense, there should be a natural association and symbolism of the different colors with the great classes of our emotional states. If man's mind is the concrete result of cycles of permanent reaction between organism and environment, then his visual sense must find its ultimate explanation in the same process, and, like it, look forward to extension and perfection on the same lines as its development has followed in the past. Now upon looking within, it is not a little startling to find the great divisions of our psychical nature corresponding with the great associations and divisions of our color-sense. It would be still more striking if we were not partially aware of the part that color has played in history and in the development of the mind. All objective existences are perhaps more vividly than in any other way represented to the imagination as colored things, and their associations with the woes and joys of life point to no fanciful symbolism, but one that is quite as real and as vital as the emotions whence he draws his mental life. Classifying the directions and methods of mental activities, we find them to fall naturally into four classes:

1. Those of the passions,—the emotions pertaining to the sensual life.
2. Those of the intellect or reason.
3. Those of utility and labor.
4. Those of the spiritual, moral, and religious nature.

These we find to correspond in an exact and specifically real sense to the four types of the chief colors previously set forth. Blood is the life, the sensual, physiological life,—the nearest, most precious, and vivid of all things or thoughts; golden light is next in its necessity and nearness to our daily life; of green we are somewhat more independent; while blue is far away and beyond the reach of our earthly cares and wants.

The symbolisms of red are, therefore, perforce, those of the two great factors of history—war and love. The passions that stir the blood and heart of men to action, the emotions of honor, vengeance, valor, love, friendship, protection, and the rest,—these are the fitting correspondencies of the rigorous challenge of red.

In the same definite way the symbols of golden light as aptly and restrictedly answer to the light of reason and of intellect which, flowing over and through all the world's ways, alone promises that clearness of vision by which we can walk in the labyrinthine maze of crowding passions, necessities, and duties.

But it is among earth's verdure that man's daily life is cast, and where he builds his home. This with its cultivation and shade, its fruitage and its various sustenance, gives him occupation, rest, food, and contentment. So in our psychological analogies, green may stand as the everyday color of labor, of use, of home-life, of peace, and rest.

Lastly, how perfectly blue represents the spiritual life of religion, of aspiration, and of morality. The impenetrable deeps of the arching sky may for a time be overcast by the passing clouds of chance and change, but the changeless blue persists, still there by day or by night, ever impersonal, and ever unattainable.

The entire history of the eye in various animals shows plainly the tentative, groping experimentation of Biologos, at one time conquered or half-conquered by a difficulty, always being changed and modified to meet emergencies, to preserve and to carry to perfection an organ without which all final causes and all ultimate purposes would be prevented of realization. Particularly suggestive is the square "about face" in changing the direction of the ocular rods and cones: in the reptilian eye they point toward the entering light; in the mammalian eye they point away from it. It seems hard to account for this, except it be on the supposition of the impossibility of making the retinal intermediate of the color-sensation as perfect as required, when the extraordinarily delicate molecular vibrations constituting its chief mechanism were located so far away from the deep-bed of capillaries of the chorioid. In the crude color-sense of the colder-blooded and inferior types of animals the molecular function could be carried on without the higher *degree* and the greater *uniformity* of heat required by the mammal and especially by man. The *natural* plan would certainly be that that was first adopted, but, according to the changed plan, the optic-nerve fibers must be divested of the white substance of Schwann, and yet preserve their complete insularity. Something over 800,000 of them, therefore, together with all the other retinal layers, must be made so faultlessly transparent as not to rob the entering and traversing light of any of its power or qualities. A profound necessity must have dictated this enormous labor, and that necessity I can imagine none other than that I have given. The fact that in death by starvation every particle of fat in the body is autodigested except that cushioning the eyeball within the orbit, also points most certainly both to the value of the eye to the organism, and to that preservation of the necessary degree and uniformity of warmth upon which our visual function depends.

Another evidence of the experimental stage and of the difficulty is the median or pineal eye of some lizards and reptiles. No longer functional it still sometimes exists more or less atrophied beneath the scales of the forehead. In the success and progress of development the Ever-Ingenious and Astute found that three eyes were not necessary, and so dispensed with one, or possibly turned the cerebral part of it to use in focalizing or utilizing magnetic currents for homing purposes.*

Here may also be noted that other great difficulty of the nutrition of highly complex and vastly delicate structures without the direct feeding of blood. Cornea, capsules, lens, and vitreous, together with a million of retinal nerve-threads and end-organs have to be nourished by lymph. Blood-corpuscles are necessarily colored, and, if present, would render any structure untransparent. Moreover, the nutrition and function of many of these structures are effected without nerve-control or aid. Biologos has to take charge of each cell again and dispense

* The explanation of the homing instinct by sight or smell is utterly disproved. Dogs chloroformed and taken hundreds of miles in round-about ways return cross-country straight home. Pigeons taken in covered baskets hundreds of miles into a new country return as direct. I have heard of one man with a tumor of the pineal gland who showed in life the most morbid restlessness and desire to walk all the time. One swallow, of course, does not make a summer.

with adventitious helps and the subordinate automatonizations of government.

The fact is undoubtedly connected with the formation of senile cataract and is the sole cause of presbyopia, that frightful anomaly of the organism, in which a most important function fails when life is only half lived.

The important fact is also to be considered that all these tremendous difficulties and labors are by no means yet surmounted. Possibly greater are yet to come. Every other organ and function of the body, relatively speaking, shows completion and perfection, but civilization is now bringing problems to the ocular divinity, and likewise to the ophthalmologist, of the greatest complexity and danger. The chief of these arise from the fact that, far more than any other function, that of vision is most important, most exercised, most bound up with every other act and with every other cerebral center and function. Considering the enormous delicacy of the forces utilized and the corresponding refinements and subtilties of the mechanism, we are struck with the fact, first, that civilization infinitely multiplies the mere physiological labor of the organs of vision by the simple increase that follows urban life, printing, writing, commercialism, handiwork, art-work, mechanics, and science.

But the chief relevant point pertains to the salient characteristic of all this increased ocular labor; its execution at what the oculists call near-range. The preponderant amount of vision in all animals and in uncivilized man was of distant objects—I mean objects beyond sixteen inches from the eye. The ability to accurately discern objects within this distance ("accommodation") was a temporary and comparatively unimportant function, and its organ was therefore one fitted for temporary use. Distant vision was the standard, and even now that is performed by the normal eye at perfect rest. Anything nearer than the horizon is by such an eye seen by means of the innervation and contraction of a delicate little sphincter muscle acting in a peculiarly "make-shift," anomalous, almost awkward, manner. All animals and savages are habitually, and all normal civilized children are congenitally, "far-sighted," or hypermetropic. But lo! within a century, after millions of years of preparation of the eye for a specific function, here at a bound comes civilization, absolutely reversing the function of developmental history, and demanding that the standard and habitual vision shall be within a foot or little more, and that the distant vision shall be exceptional. Moreover, it demands that both distant and near vision shall be equally perfect vision. We are witnesses of the attempt to develop a new muscle in order to meet this emergency,—the Müller Ring muscle in hyperopes—acting in an entirely different manner from the ciliary muscle, but to the same end and complementing its insufficiency. If the suddenness and, as one may say, the virulence of the on-coming of civilization had not been so great, and if an artificial help had not been at hand, the attempt would probably be successful. This may now be doubted. But happily with the hurt, in the other hand civilization brings the healing, and though spectacles are a "poor excuse" they must be used by civilization's servants a thousand times more than in the past or present. The eyeball is a soft rubber-ball-like structure, any slightest departure of which, caused by many conditions and circumstances, from the most ideal and accurate shape—in length (myopia) shortness (hyperopia), or curvature (astigmatism), at once results in dangerous imperfections of vision, and disease of itself, of adjacent, or of distant organs.

To such facts must be added others that are noteworthy, *e. g.*, that the position of the eyes must be in the most exposed part of the organism; that, by the necessity of their free motility, they must be in a condition of semi-detachment from it; and that, by the necessity of quick directableness toward different points of the compass, they must be moved by twelve tiny muscles all in perfect balance and counterbalance. The least inco-ordination in power or in function of these muscles, may at once breed trouble.

One could go on indefinitely detailing the dangers, the difficulties, and the delicacies of this organ and of its Artificer. At the risk of wearying the reader I shall further mention but two: The usual reflex nerve action from an irritated or diseased organ normally brings the reply of inflammation, reaction, motion, etc., back to the organ irritated or affected. The out-valuing importance of the eye over most every other organ makes this a dangerous result, and hence an exception is instituted to this otherwise almost invariable physiological rule. The result is that the pain and the disease-effects of abnormal ocular function are felt elsewhere than in the eye (reflex neuroses) and usually as pain or ache in the brain. This morbidity and anomalousness of reflex-return is doubled by the fact that facial beauty is dependent upon, almost consists in, the limpid, soul-expressive beauty of the eye. Now the influence of beauty in sexual choice and its powerful rôle in biological history are plain to the Darwinian and to every one who has never even heard of Darwin. As the natural return of the reflex to the eye would quickly result in inflammations temporarily or permanently destructive of ocular and so of all beauty, the subtle wisdom, nowhere else so severely tested and so evidently present as in the eye, stores up the irritation in the brain cells (headache), or shunts them elsewhere. Not one patient in a hundred complains to the ophthalmic surgeon of pain or visible disease of the eye. The more unrelieved the "eye-strain," the greater the production of irritational unused stimuli, and the greater their flow or overflow to other centers. The more and the longer the storing of the continuously generated steam, the greater the pressure, and the more harmful and certain the release when a weak rivet or other opportunity is present. A natural and a customary method of obviating the injury in the mechanism considered, is to shut off the fuel and the firing by means of anorexia, digestional interference, or some other disturbance of function.

A last illustration is that of the psychological influence of difficult ocular function. Thousands of children are to-day being switched into less intellectual lives, are being restricted in mental development or occupation in life, by the fact, to them unconscious, of irksomeness and of discomfort in reading, in study, or in other intellectual task that necessitates labor at near range with an organ illy adapted for the work. Activity therefore finds outlet in sports, or in other physical methods and occupations; if the task be forced, worse evils ensue, and in either case parent and child both suffer.

Of profound interest to the philosopher is the embryological fact that the essential organ of the eye, the retina, is a cerebral outgrowth, and not a dermal structure. *The brain comes out to see.* It is not the light, not the skin, that goes in to stimulate and interpret. The doctrine that an organ and an organism is solely a mechanically rigid reaction to the environment is disproved by the developmental history of every organ, but notably and beautifully so in this most psychical and most useful of all the senses.

I cannot forbear reference to a possible artistic extension of the function of vision similar to that already made actual in hearing. It is natural that music should have been born and perfected before the as-yet-unnamed art of visual or scenic opera and representation. The lower and simpler must precede the higher and more complex, and the function of music is distinctly emotional. But intellect is higher than emotion, and vision is *par excellence* the sense of the intellect, at once its servant and lover. Moreover, a series of symbols lie at the hand of the visual artist, of far greater exactness and specificity than Beethoven or Wagner ever had. The glory of Wagner will be that he sought, and successfully too, to supply the want of definite musical symbols and typifications of concentrated emotional experience by a somewhat artificial but withal significant series of representative harmonies. He tried to mentalize emotions, to classify them, and to epitomize and symbolize them by the *motiv*. He bravely and creatively marched into the unknown and definitized his art with this *de novo* production, which has revolutionized music and that will set it in progressive advance to yet undreamed-of vigor and splendor. The coming opera of visual harmonies and delights is only suggested by the perfection of scenery of the modern stage, approximated by the Berlin "Urania," but no one has thought of making the eye the sole organ appealed to in a great work of creative art. This combination of sculpture and painting; the symbolisms of color; the scenes of landscape; of water, still, rippled, or colored; of multi-colored illumination; of atmospheric effects; of sunsets,—these form an alphabet of visual elements wherewith an artist could make an evening of pure, peculiar, and of more intellectual enjoyment than has yet been known. Nature rarely unites all the conditions of a perfect landscape or sunset. For example, it is only once in the life of but few that one stands upon a plank or two in a still, horizonless ocean, with nothing for the eye to rest upon, a cloud of fog distantly hiding everything above, below, and all about in a monotone of gray, and in the east a huge ball of red fire gleaming, and in the west another similar globe of lurid light,—self lost in the midst of nothingness, the lifted hovering Ego stared at from the depths of infinite space by the ghostly glory of those two motionless, splendid, terrible, cosmic eyes.

In viewing sensation as a whole, it seems impossible to withhold allusion to the idealist's contention that because all qualities of objects, one's body, nerves, and nervous centers, just as well as the extra-physiologic world of inorganic matter, are only to be stated or thought of in terms of mental experience, therefore naught but mental experience or mentality exists. When one gets wholly within the machinery-room of subjective being, and, *à la* Berkeley, will listen or indeed can listen only to the buzz and din of one's own mental machinery, it is not difficult to delude one's self into a sort of somnambulistic daze or hypnotic stupor, and to simulate a belief that objective facts do not exist, or that they exist only in relation to sensation, and that "the organs of sense are themselves so many groups of sensible phenomena existing only in the mind, the body itself being simply a part of our mental experience." "If," says an idealist,* "if when I die no one is present or observes me, there would be no physical death, properly so-called, but simply the inexplicable fact of my ceasing to feel and think." "The whole sensible world, heavens, earth, and the physi-

* *First Steps in Philosophy*, by W. M. Salter; Kerr & Co., Chicago, 1892.

ologic body, are equally unreal if regarded as a self-subsisting thing apart from a sentient subject."

These are spoken words, but nobody ever imagined the supposed fact they represent. They are simply jugglings with words after such words have been disconnected from a vital significance and relationship with facts. No Berkeley, even of the most super-refined and gigantesque sort, could look at himself in the mirror and say that his physiologic body is non-existent when he goes to sleep, or when neither self nor any one else is present to evoke that body into an illusional objectivity by the power of "mental experience." * The first requisite of philosophic thought is to recognize differences in facts,—indeed this is the beginning and end of all intellectual function. To see no difference between the objective, self-existent causes of sensation, and the ultimate products of sensation, is simply philosophic barbarism or monomania. It is not healthy thought. To deny objectivity and self-existence because the forces and impingements of matter must be translated into sensation before being presented to mind, is to flatter one's self that dreams and imaginations are facts; it is a lazy sort of dogmatism that contents itself with the whirr of subjective machinery, rather than look with virile earnestness into the question of whence the forces, the coal, the steam, and the piston, that propel the whole. The natural order of true scientific thought consists first in the study of molecular physics, or the nature of the objective forces; secondly the nature of the receiving sense-organs is considered; then follow the problems of nerve-force and nerve-transmission, and the nature of

* I have a little nephew who, in his grief and between his sobs that his mother had gone away, piteously complained that she had not at least left him her body while she was away. True philosopher, that baby!

sensation proper, together with that of the psychic reality; finally is reached the consideration of the whole as regards the elements of product or reality that are contributed by each factor.

Sensation of any kind and as a whole is but the mechanism of that differentiation of function whereby the purposes of Biologos are more easily and perfectly carried out. We are obliged to admit that the undifferentiated cell has all the same kinds of powers and purposes as the highest and most specialized cell or being, the sole difference being one of degree. Thus as to sensation before being assigned a special function, the living cell shows reactions to light, to air-waves, to pressure, to heat, to cold, to odors, to poisons, to anesthetics—in a word to all physical and chemical agents and agencies. It also shows every physiological function possessed by the highest organism—that is, contractility or spontaneous change of form; irritability, or response to stimulation; respiration; anabolism, or the indrawing of new material, the building up of greater complexity, and the repair of waste; katabolism, or excretion; and reproduction. The lowest uses of sensation are the nutritional and protective, which are sufficiently evident, but the higher are not so plain and may be described as threefold: (1) to gain control of molar motion; (2) to learn the qualities, actions, and laws of the inorganic world; and (3) to utilize acquired power of molar motion and knowledge of matter for spiritual uses, enjoyments, and progress. The enjoyment is largely summarized in art and esthetics; the progress in the increasing dominion of mind over matter, or the control and ordering of the inorganic substances and forces of the universe. The final cause of the entire nervous system and of subordinate functions is to be found in that aspect of differentiation of function consisting in the more perfect control and use of an order of highly complex cerebral cells, more accurately and thoroughly responsive and obedient to the control of Biologos, and whereby He can more perfectly effect His purpose and make permanent His use of the whole organism. Nor should we forget that in His specialization of control-cells these immediate agents of the highest incarnation are also the final end-products of organization, and that they have the exalted function of carrying to Biologos the synthetized quintessence of the inorganic world, and the revelation to the mind of the sublime products of objective life. They are at once the first deputies of mental control and the last finishers of received influences, the final co-ordinators of concentrated and representative experience. This is true, because the function of cell-pliancy and the perfection of metaphysical control must be as much or more a matter of progress and of differentiation as that of musculation, or of secretion. God's control of the process of incarnation, and the objectivation of Himself is progressive, and these higher stages are dependent upon the delicate poise of temperature, the permanent supply of ever-higher orders of nutrition, and upon the ever-increasing complexity of cerebral cell-life and function. The struggle for existence has largely passed the nutritional stage, and has merged itself into a struggle for intellect and art; and a similar change has been reached in the task of incarnation, which is becoming one of an ever-higher differentiation of cerebral cell-systems, based upon the systematization and automatonization of subordinate centers and adjusting mechanisms of similarly increasing complexity. And, just as by artistic creations and exercises in music and poetry, we make application to esthetic uses of the mechanisms born of utility, so, having conquered and learned the world, we may suppose that God is preparing arts unknown to be played with the completed products of incarnation. And as man is the highest completed product of incarnation, so, as we learn loyalty, we may expect each personality to form a line or a thought in the epic of the coming poem, or a tone in the approaching symphony of humanity.

CHAPTER VI.

"EVOLUTION."

"EVOLUTION" is the word most drummed into the ears of the young philosopher and the godless child. I have some difficulty in controlling my amused contempt at the mere mention of it. The word is, consciously or unconsciously, the evidence either of disingenuousness of heart or of indiscrimination of intellect, and usually of both combined. By its devisers and users the term is applied to two unrelated and unrelatable orders of phenomena, the living and the non-living. If it be applied to both it must denote only a colorless meaning, like that of the word, being, or fact, or phenomenon. If it can specify any quality of the one, or any considerable peculiarity beyond the mere fact of existence, it cannot specify that quality of the other, because matter and life have little more in common than mere existence. In the entire realm of inorganic matter untouched by life, all changes are non-purposive, proceed purely mechanically, repeat themselves aimlessly, and lead nowhither. In other words, there is change but not progress, causation without object, rigid uniformity but not freedom. It is perfectly logical that after the designation of two such diverse types of phenomena with one name, its shrewd fathers should proceed in the natural course of self-stultification by denying the existence of life, purpose, and freedom in biological phenomena. To them the mechanical clank,

clank, of a blessed determinism is both natural and certain. But their neophytes and pupils will by no means so see the world; they protest that there are bird-songs, mother-love, laughter; that there are sports of children and "sports" of morphology (how the last must be to the scientist as holy-water to the stage-mephisto,)—and if "evolution" covers these things, the sensible child will unconsciously say, "Why, then it can't be such a horrid bugaboo as our teachers would have us believe." There is therefore a lazy, cunning, double sin thus dextrously allowed and encouraged by those who should have both more honor and more intellect. Like a great many more such words "Evolution" is a large garment for covering a multitude of sins. It is about on a par with the indirectness wrapped up in the word agnosticism, or the words God, and Father, as used by some "liberals" who in their hearts acknowledge neither personality nor guiding love in their brand new mechanic-god, stuffed into the old wine-bottles of an ancient faith. "Agnostic" and "evolution" thus connote meanings very different from those held by their concoctors. The ethical sin consists of the concurrence in such connotation, and in giving out that ticketing a process as evolutional explains it, or does away with an intelligent evolutor. Thousands who make a God of evolution do it by stealing attributes from the old divinity and hiding them carefully under the cloak of the new one. The new cathedral is made of stones quarried from the deserted ruins of the old temple; the name of the deity is changed, but the worship secretly continues.

Take the official definition: "An integration of matter and concomitant dissipation of motion; during which the matter passes from an indefinite, incoherent homogeneity to a definite, coherent heterogeneity; and during which the retained motion undergoes a parallel transformation." With perfect candor one may first ask, has one learned from it anything of the world? It is so colorless and vague, so powerless to help the imagination, that one might quite as well say, things change and multiplicity results. Even if true, it is so indefinite as to be of little use to the inquiring mind. The generalization does not even glitter, so devoid is it of specificity and life.

Even if true of the inorganic changes, is it true of biological phenomena? Certainly a large portion of vital phenomena illustrates the reverse of the theorem. One full half of every living metabolic process is katabolic, and katabolism is the exact opposite of the definition. Indeed, according to Mr. Spencer himself, "reverse distribution" is precisely one half of the world-process organic or inorganic, and "reverse distribution" or "dissolution" yet awaits its Spencer to base a universal philosophy upon a neglected half of all phenomena. The characteristic of organic "evolution," *i.e.*, of living anabolism or constructive change, is progress; that of all inorganic "evolution," however, is simply change, change by pure mechanical forces, undesigned and leading nowhither. To cover both processes by the one word tells little enough about physics, but it tells absolutely nothing about the distinctive quality of metaphysics, unless it be that, temporarily, perhaps, unity results in diversity—a somewhat unsatisfying creed. But the other half of the truth is the reverse, and to say that diversity results in unity is precisely as true a statement, and often more illuminating to the mind. The wonderful unity produced by ten thousand bees, their exquisite workings, adjustments, divisions of labor, sacrifices, interdependencies, and ingenuities, produce an astounding singleness of result. So with every illustration of mentality in nature. It takes billions of separate cells, "diversities," to create the unity of each organ and of each organism.

As applied to the inorganic world, evolution can mean nothing more than the purposeless sequence of mechanically caused change. There is not, legitimately, any "development" here, because that word connotes methodical progress and precedent purpose, and these are characteristics that apply only to biological phenomena, whilst the great gloomy and mysterious word Evolution is kept as a sort of primary Fate that rules all worlds, things, men, and gods.

Limited to its strict etymological signification and stripped of unlicensed connotation, the word evolution can only be applied to one aspect of one half (anabolism) of the biologic process, whilst it fails to cover many aspects even of anabolism, and has no application whatever to katabolic changes whether physical or metaphysical. The one idea clearly called up in the mind by the word is that of a ceaseless process, of which every step, stage, or instant is derived from that of the immediately preceding condition, by the absolutely necessary (mechanical, deterministic, or unavoidable) action of natural (or mechanical) inherent forces. Now this, as regards living things, is utterly false. Every cell and somacule of living matter is what it is by the incoming, re-creating, and in-living control of an extra-cellular intelligent force. This scorned fact is deftly slipped out of sight by the further explanation on the part of the "evolutionists" that the living organism is distinguished by a rigid (mechanical or necessary) adaptation of the organism to external circumstance, or reaction to external changes. Here again is a double untruth, by confusing and by ignoring. The confusion consists of making "adaptation" and "reaction" necessary, unexceptional, and mechanical, whilst to all of us simple folk, the words unavoidably imply intelligence, discrimination, mentality. A "reaction" that does not flow from mentality would speedily end in the death of the reactor. The ignoring consists in not seeing that by all odds the larger number of acts of living beings have not their origin or anything whatever to do with "reactions to external changes and stimuli." All growth, as of the ovum and child, proceeds from within; the rôles of the reproductive instinct, of hunger, and nutrition, are called forth by no external circumstance but spontaneously arise within, always dominating and using external circumstance. What have music, art, aspiration, and moral ideals, to do with external stimuli? Verily it is pathetically strange that a mind so great and calm as that of Spencer could be dominated by a preconceived theory so much as to ignore and misconstrue the greater part and the most striking of facts. It shows the fatuity of deductive reasoning; but sadder still, it shows how a captivating theory may enslave the mind of a supposably inductive thinker, and, bewitching thousands of less discriminating followers, handicap science and lead modern thought into an *impasse* of error.

Leaving out of consideration the entire domain of the organic and metaphysical, we find that the "laws" of the atomicity of matter, and of the persistence and intertransformation of forces, explain all the meaningless changes occurring in this world. Turning to the world of living things, let us grasp the essential idea or mystery desired to be expressed, or mistakenly understood, by the word evolution. I mean the mystery of the gradual process. Here is something every one sees is different from any purely mechanical process. The scientist's sin is in ignoring, jumbling, or denying the difference, whilst the waiting world stands dazed before the inexplicability of it. As to absolute origin I have of course no explanation to offer except that already emphasized and to me very satisfying one, of its creation and instant in-living by the self-incarnation of Biologos. But the very heart of the mystery yet untouched is the chronicity or slow progressiveness of the process, the facts of rejuvenescence, senescence, and death of individual forms, the continuity of special character and purpose throughout any one line of these sequent organisms, and the domination of one harmonic character and purpose over and through the entire order of all sequences. The explanation of this I have never seen attempted or known to be thought of as within the scope of human reason. And yet the key seems to me to lie placed in our hands by the most childishly easy and simple perception of the controlling necessity or struggle that constitutes the essential fact of the life-history of every cell, organ, and organism, of one's own life, of all history, vegetable, animal, or human. I mean nutrition, in its full significance, as the condition of cell-building and cell-control, cells being the primary intermediates or elements of incarnation, through the association and conjoined work of which in organs and organisms, Biologos is with difficulty working out His ulterior purposes. The condition or difficulty is plainly nutritional, disease being impaired nutrition; death of the individual organism being the seeming victory of the difficulty; sociological progress, ethics, the good of civilization, being firstly and physiologically the sharing, assuring, and stabilizing of nutritional success, and, secondly, the working out of the metaphysiological purpose. The life of plant, animal, or man, from seed or ovum to death, is based upon the necessity of gathering food from without. Disease and death are solely the consequences of incapacity to gather or to utilize the food. The condition of the cell or of the individual (organized family of cells) is the condition imposed upon God, literally and in large; and the great mysteries and characteristics of organic history, viewed in their physiological and strictly biological aspects, are individually and collectively the results of this difficulty or condition pressed upon Deity by the nature and laws of matter. Blinded by an illogical deductive habit of mind, controlled by an unscientific scorn and a " reaction " from religious habits, Science has failed to realize the significance of the plainest of all facts, which is mentality in living things put to enormous labor by the difficulties of nutrition. In a world of mechanics, with so great differences and fluctuations of temperature, with storm and drouth, with flood, ice-ages, submergences of continents, volcano and earthquake, and a hundred like enemies always about and before, Biologos has had to take precautions, forefend dangers, devise mechanisms of defence and safety, preserve balances and control, hold by infinite devices and watchfulness against the elusive slipping to rest or out of control of the dead atoms,—all in myriad ways, and with a subtle wisdom, of which the mind of man has scarcely begun to find the faintest clue or hint. And with it all He has had to keep the march onward and upward towards the final result, civilization, and of what civilization has scarcely yet commenced to prophesy. Because, so far, all, or nearly all, is but prophesy and preparation. Herein rests the blessed logic of faith. ' Though he slay me, yet will I trust Him." But this faith must be supplemented by an intellectual perception of present difficulties being conquered, of progress going on, and of final victory certain, whence will arise in the heart of mankind a living intellectualized religion, and a large reasoned content that that heart has never yet known. And such an intellectual religion will inspire men to become God's co-workers and co-partners,—true sons of God, according to the grand old promise,—whereby "the coming of the Kingdom" will be immeasurably hastened, and wherein, also, that Kingdom really consists.

One suggested thought may appropriately be interjected in this connection—the instinctive clinging to life on the part of cell and organism. In part this may be interpreted as a corollary of the automatonization of function, the persistence in action of specialized cells. But this by no means accounts for the "never-give-up" of the non-self-conscious organism, the acceptance of life when its meaning is utterly lost in denied and distorted function, in degradations of a thousand types and degrees, in dwarfings, stuntings, and reversions, and in most pitiable adaptations to tyrant condition. There is no suicide among God's creatures who have not been given deputed power, *i. e.*, it exists nowhere except in disloyal man. His true lieges dispute the last inch of place gained, and die only when killed. In part again this is explained by the positive truth that upon the brave fighting for and holding on to life on the part of the humblest plant or animal may depend the fate of the entire fauna and flora of a continent or of a world, threatened as it is with wreck by the possible disasters and certain changes of a physical world such as ours. But even this would explain only the aspect of positive assertion, the exuberant energy to extend; it would not explain the instinctive heart-clutching fear of death and the acceptance of life under conditions that make it absurdly illogical. That, I take it, is explained only by the terrible struggle Biologos has had to get his foothold in matter, only by the memory of the long and bitter labor, and of how often defeat may have really happened, and of how much more often it was averted by infinite device, resolution, and exertion. The fact that the labor of the primary step of incarnation was a herculean task, that it was at once a subjection of obstinate enemies, and a watchfulness of surrounding enemies still more uncompromising—a true invasion into a hostile world,—gleams out upon us from every biological fact. And chiefly from the astonishing observation that Biologos always proceeds from his foothold gained, and never from a *de novo* beginning. However useless and degraded and lowly the speck of living plasm, it is the vantage-ground of intrenched victory,—nay, of a hundred victories over the stubbornest of enemies, and *no surrender* is its sole battle-cry. From this citadel and "base of supplies" the army may go forth to the conquering of a world. This says plainly enough: "No new attempts or experiments. Like the telegraphic current of electricity, we can proceed eastward around the whole world and reach yonder western shore rather than jump this little space." Spontaneous generation and the creation of the homunculus will not take place in the laboratories of the nineteenth or of the twentieth century. No scientist ever saw the first stirring of life in dead matter. *Omne vivum ex vivo* implies that a difficult task once completed will not be recommenced for the sake of curiosity or gymnastics.

And the difficulty of the task is the *raison d'être* of "evolution," defined as the gradual unfolding and development of life on the earth. The only way to reach man was by means of millions of years of endeavor and through the millions of progressively preparative stages of precedent vegetable and animal types. There is a tendency nowadays to sneer at the essential truth of the development theory. "The theologians are now taking courage," it is said, and, secure within their fortress, the pickets of the scientists are indifferent to the bold sorties of their enemies. Both furnish diversion for the non-combatant spectator, who if the scientists were in much danger from this little attack would whistle the signal of alarm that would put the special creationists to easy rout. Missing links, and all such, are only needed by the minds whose processes, like bad machines, move only with mechanical clank, clank. The slightest study of our world proves that only from the present $\pi o\tilde{u}\ \sigma\tau\omega$ does the hidden life proceed to develop distant types through intermediate modifications and correlated stages. Backward only do the lines lead to unity; forward no type fuses; that way is everlasting diversity. "Phylogeny is the repetition of Ontogeny" is a maxim based on inexpugnable fact.

Mr. Spencer's unworthy and fallacious attempt to smudge over the marks of difference between phenomena that are purely mechanical and those that are meta-mechanical, however necessary to his theory, is good intellect wasted in a bad cause. A critic has justly said that even Mr. Spencer's theory demanded a beginning of life on earth. The early heat left no room for doubt about that. Mr. Spencer's answer consists in placing the prestidigitator's *foulard* of infinite time and infinite numbers of progressive stages, over the trick of creation. When the delusive mysterious silk handkerchief is deftly removed, the completed product of living form is revealed to the gaping crowd. It would have been easy for some "logic coach" to have disclosed the stupid fallacy of Zeno's paradox, even were it not *de facto* sufficiently evident that the hare would catch the tortoise, if, peradventure, he were silly enough to enter the race. The subjective possibility of imagining an infinite number of fractions between one and two, and the supposed consequent impossibility of uniting two integers, does not weaken the truth that one and one do make two, nor do the infinite number of the stages imaginable between mechanical and vital phenomena do away with the fact that the two orders are for evermore distinct and nonidentifiable, and that they always proceed from distinct, unidentifiable origins. The lowest living cell and the highest non-living molecule are separated by a chasm so wide that no thought has ever legitimately bridged it, and so deep that no sounding has ever fathomed it.

There is one phase of the relation of the cell to evolution to which I have not known allusion made. It is admitted that "phylogeny repeats ontogeny," or that the life history of the individual is that of the race, or *vice versa*. Now every organic individual begins as a simple cell, and therefore the first living matter on the globe must have been that of a single cell. The primal and typical mechanism of the incarnation of life was and still remains cellular. There is no truth of science more absolutely certain than that summed up in the maxim, *omnis cellula e cellula*. Virchow has conclusively shown that no cell is formed anew, *i. e.*, out of or by a supposed plastic lymph. Every cell springs from a previous cell by heredity and continuity. There is no epigenesis or *generatio equivoca*. Biologos can only work from the existing mechanism of cell-formation, from the secured foothold, outward and onward. The power to hold vast numbers of subordinate systems of complex molecules about a common center is possible to the cell-mechanician only with and through the great complex that has been once established. There is no possibility of nutrition and further outbuilding of cell-life otherwise than by means of the established mechanism. If one does not perceive a difference in kind between this cell-genesis and cell-nutrition on the one hand, and the formation of a crystal, then it is useless to argue with him. The term *organic crystallization* is an egregious contradiction in itself. It is a myth of the illogician and of the materialist. The formation of the crystal is the result of purely physical forces of attraction and repulsion. Cell-genesis is a true birth following growth, guided by a power using as tools all physical forces of attraction and repulsion.

CHAPTER VII.

REPRODUCTION.

A WISE man once said that whilst great philosophers were learnedly speculating about the world, hunger and love were producing and controlling it. It is a profound truth. The rôle of hunger, as we have seen, reaches down to the very essence of the process of incarnation. To establish the stable equilibrium of a complex molecular system by bringing into it subordinated molecules, and then exhausting them of their contained force, this is the first and continuous work of every living cell or rather cell-master. The problem of nutrition precedes, underlies, and controls every other biological problem.

And even also that of sexual reproduction. The wise man might have said that hunger alone ruled the world, if he had seen what a spiritual-minded science has at last begun to get glimpses of, viz., that the entire mechanism of reproduction, together with all the influences and consequences of sex and sexualism, are but factors and devices of Biologos to overcome difficulties of nutrition and to turn these difficulties into the very means of progress. The last word of a careful and enlightened science (to which result it is highly suggestive to note that the unprofessional, the enthused amateur has contributed the most living results and done the most patient work) is that, from the lowest flagellate to homo the function of reproduction is to bring about rejuvenescence, or, what is the same thing, to avoid senescence. Weissmann's theories, and all the rest of the materialist and mechanical "explanations" of heredity, are found unscientific, and science is driven back straight to the acknowledgment of " conjugation " as a device to avoid deterioration and running-down of the clock. This rejuvenescence of the running-down or exhausting cell is effected in three ways: 1. By Rest, whereby the innate powers of recuperation, the healing power inherently possessed by the cell, restore and restock it with new vitality; *i.e.*, it takes time to carry out anabolism even in the presence of an existing food-supply. 2. By a change of a mode of life (as in parasitic fungi), whereby the cell or organism, again without division or conjugation, recuperates itself in new conditions. These new conditions can be thought of in no other way than as those aiding toward improved nutrition: it seeks new pastures. 3. By conjugation and fertilization, *i.e.*, by the fusion of complementing organisms, that supplement each other's deficiencies, and double the life-forces of the single and separate *gamete*. Scientifically, therefore, the origin of sex consists in the gradual differentiation of *gametes* (uniting cells) into categories of distinct size and habit, and the reunion of the diverse (male and female) cells, each with diverse histories, deficiencies, and excellencies, reinvigorates the resultant organism by neutralizing deficiency and reinforcing advantages. Whatever temporary exception seems to break the force of the theory, endogamy (in-and-in breeding) must finally prove bad for the type, and exogamy be to its advantage. From all this is plainly apparent the controlling influence of the cellular and organismal struggle for nutrition, and against the running-down of the clock-work of the physiological system or unit. It presupposes and rests upon the conception of a cytology that views the cell-system as made and upheld by an extra-cellular master, who is at once steam and engineer, keeping the machine at work by ever watchful energy, and performing the task, impossible to the pure mechanic and materialist, of producing from two of his like-unlike mechanisms a newer and better "rejuvenescent" one, to replace the elder when these have been "worn out." The most ludicrous fancy of a baby locomotive, spontaneously growing and born from two larger engines, is quite as sensible as the materialistic world-conceptions of some "philosophers," and gives one a realizing sense of the difference between a living and a dead organism.

But I wish to pierce still deeper into the heart of the mystery of reproduction. The word, rejuvenescence, does not still quite reveal it. The labor of reproduction, viewed as a cosmical fact, is plainly a tremendously expensive and difficult piece of work; it is clearly an ingenious and roundabout way of obviating a worse result, and it is clearly imposed by the material conditions of incarnation. Like all things material the cell is liable to "wear and tear." With all His incomprehensible ingenuity, Biologos has been unable to construct a permanently durable cell or organismal home. The unstable equilibrium with which He is able to endow the cell lasts but a little time, and the most noteworthy peculiarity that I find in this world is that the new house that Biologos makes before the old has become untenable, and into which he moves or transplants Himself, is built precisely on the model of the old ones. The ground plan and essential superstructure are never changed except imperceptibly or slowly, and there is seldom much sudden change in the unimportant details of ornamentation, etc. There is slow change and progress everywhere. This is the "law" of the development by types, and extension only along the lines leading out from all the past. That each animal breeds its kind is so uniform a fact, that its strangeness strikes no one; and yet, exceedingly strange it is, and requires explanation. An architect that makes his houses all alike is called stupid and unimaginative. Now the Divine Architect is far from stupid or unimaginative. Why, in the quite infinite spontaneousness of his imagination and power, does he not create an infinite diversity of forms, not on type-lines, but giving exuberant play to diversity within types, and at will jumping the boundaries of genus and family, produce from one set of conjugating reproductive cells, beings utterly unlike the parents? To put the question thus is to introduce the *reductio ad absurdum* that answers itself. Such would certainly be a world far inferior in order and interest to the one before us. Besides, in the multiplicity of types quite beyond enumeration, each following its own line of development, we have precisely the variegated world we desire. But the deeper reasons are plainly those of the nutritional necessities of his organisms once that they have been created, whilst still deeper lies the fundamental reason that the biologic Architect has to make not only his own tools but his own material mechanisms, and that from an existing organism can be obtained only germs of like chemically-constructed somacules and cells. The new house grows out of the old, and in essentials and within limits must be like it.

Digression passed, we return to the fact that the cell-system, as well as the organismal unity, can preserve continuity of function for but a little while. There is something in the deterioration that suggests the simile of the clock, not only needing both constant and periodical "winding-up," but, despite all, finally wearing out. For a time the cell-units as they die may be supplied by new ones, and the organismal unity still preserved, but in time the larger unit goes inevitably through senescence to death. The very nature of life, and the desperate devices taken to circumvent it, show that death is not a part of the wish and plan of Biologos. It is clearly the hated condition imposed upon The Deathless One by His temporary servant, matter, and with all His ingenuity He has not yet learned to conquer it. In reproduction He compromises and circumvents; it is not a "victory in open field," but is, as it were, a *ruse de guerre* against an impregnably fortressed enemy. I am not at all sure but that with permanency of condition, uniformity of nutrition, and temperature, the planet made homelike in all its parts, and God aided by man's intelligent loyalty, human death may not be almost or quite conquered. It is certain that our lives might be very much lengthened, and all that is really desirable as to earthly immortality fully realized. With some of these conditions attained the tree has learned to preserve its beautiful life for thousands of years. Perhaps the time has not yet arrived, but some time the scientific study of the conditions and means of preventing senescence and death will be of profound importance and interest. Of course, the avoidance of harmful "wear and tear," by hygienic measures will be necessary, but the encouragement of the unconscious vital nutritive process that can replace old cells with new ones will constitute the primary and principal means until the true immortality of the cell itself shall have been secured by the renovation of its parts. Not until railroads and steamships had been perfected have the facts of certainty and uniformity in the supply of the crude materials of nutrition become secure, and even now for but a limited number, so that the struggle for temporary maintenance of existence and reproduction was dominant, and Biologos and His creatures could give little or no attention to the indefinite lengthening of individual lives. With civilization or nutrition secured, the crusade against death itself may be, and even already has been entered upon. The shameless Brown-Séquard savagery teaches many truths wise and otherwise, but it at least showed the hope and the pathetic appeal to Life in the denier of life. If ever gained, the victory will, of course, not be sudden: God's victories are never so; there may be cycle-long approaches and temporary failures, and death will hardly be appeased by continuous death-sacrifices to him. Life loves all her children, not man alone. Man is hardly cunning enough to play tricks with God.

In the meantime it is consoling to notice the inexhaustible kindness and thoughtfulness of God in breaking the force of all penalties and tragedies. Nothing so arouses love and gratitude to Him as vision of the subtle foresight with which He has softened all sorrows and provided compensations for all those who are loyal to Him. The greatest tragedy in the world is death, and there is no mitigation of the force of the blow in case of unmerited and early death. But that is not necessary, nor is it in His plan. We all know that the large numbers of those that die young are solely due to the sin, perversity, and the disloyalty of men. God is only waiting for help, waiting for us to act up to the light we have already gained. But the death of those who have lived out the "allotted term" honorably and bravely, is to self and loving ones, like sleep at the tired day's ending,—none questions that it is fitting and right.

The next greatest tragedy is the existence of the "defective classes." Here again God is waiting for us; these are of our making. When we are determined, they may be prevented from coming into existence. The diseases that cause so many, are preventable diseases, and the social sins that cause so many more are likewise avoidable. The hopelessly idiotic should be legally and kindly killed. The insane may be made comparatively happy and inexpensive by the wise kindness of their guided labor; the blind and the deaf and dumb may be educated, made happy, and self-supporting with a little kind judgment and love upon our part.

And while God waits upon us He is not forgetful even of the slightest touches of affection, but is ever watchful, the dear Father! to heal the physical hurts, to devise sweet compensations and rests, guarding precipice-edges with roses, or with thorns if we are heedless, with death if foolish; suggesting new outlets and ways, hiding past griefs in present joys, and above all rewarding those who love Him with blessings others cannot know or feel. All good literature is certification of these facts, and in all our lives are they exemplified each hour of every day.

But in three things especially has the divine kindness shown itself most surpassingly: these are the beauty of woman, the exquisiteness of love, and the sweetness of child-life; their presence and influence in our life are the chief of its attractions, and the dominant notes of its music. Each has a threefold glowing significance—each is the exultant proof of victory over death, the smile of satisfaction reflected from the very face of God as He beholds that His world still lives; each is the divine benediction upon His children who have been obedient and loyal, because they are rewards given only to the good; and, finally, each is but a glimpse and promise of the richness of the divine life, waiting only opportunity to be revealed, and eyes appreciative of the vision. Because, if such wonders are shown as the mere certificate and proof of a difficulty only partially conquered, what must be the unshown wonders awaiting revelation, when all difficulties are fully overcome, and there is the unhindered way to realization of ideals, hitherto postponed by the struggle with fateful circumstance and obstinate material?

Buddhistic and Christian asceticism, and, whether religious, philosophic, or sensualistic, pessimism generally, has represented the beauty of woman and sexual love as the baiting of the devil's hook. With unexampled clearness and splendid analysis the great Schopenhauer has set forth this view, and if he had but put God in place of his diabolic will (blind, and yet, illogically enough, superbly, even fiendishly, cunning), the exposition would have stood as a marvel of physiologico-philosophic reasoning and description. It is perfectly useless and philosophically wrong to blink or ignore the evident partial truth of this view. From the Iliad of Homer to the Iliad of to-day's suicide the fact is exemplified, and the life of every one of us has been either moulded or deeply influenced, directly or indirectly, positively or negatively, by it. It is the constant theme of all literature, the staple subject of all joke, the secret and invisible hand ever leading to all kinds of fortune or of misfortune. The philosophy that leaves such an agent unobserved and unexplained is simply no philosophy; and the theology that leaves out such a plain evidence of supernaturalism is no theology. The majority of philosophies and theologies may indeed be quite accurately described as characterized by their failure of interest in what God is most interested in. Their concern is in inverse ratio to His.

To every adult it is perfectly evident and clear, this purpose of the sexual instinct and mechanism. But with the philosopher many questions arise: Is the entire mechanism the best that could be devised? Are the evils resulting from its working chargeable to God or man? Do the good results more than balance the evil results? What about the future? According to his standpoint and the clearness of his mind, each person's answer to such questions will be different. My own would be something as follows: The mechanism is of course the direct work of Biologos, essentially the same in plant, animal, and man, the plain method of the intermediation of reproduction. Reproduction in its entirety is a device to circumvent death, and all indirect means of reaching a distant end, all obviation of difficulties, are apt to show the machinery quite too plainly, and reveal the quality of the "makeshift." Now I have not the faintest desire to minimize the truth of the "bait-and-hook philosophy"— neither must it be exaggerated. It must simply be recognized and reasoned about with entire calmness and frankness, and finally utilized—improved, if you please, for the evident ends, and according to God's plain directions. It is an infinitely serious matter, despite the knowing "wink" and leering smile. When one comes to ask the cunning winker in what his self-satisfied secret knowledge consists, the "open secret" is found to be solely a 'cute little trick of stealing the bait and escaping the hook. But if God is fishing for fools, as the *roués* and bachelors grin, the most perfect specimens are the grinning bait-stealers themselves, because, if the bad metaphor be allowed, the bait without the hook is poison and death. And to folly is always added wickedness, in the inevitable suffering and degradation of the victims.

The problem for God, for man, for progressive humanity, and for the philosopher, is the most perplexing and dangerous. I judge it to be the crucial difficulty in humanity's future. Civilization is more endangered in its solution than by plutocracy, war, anarchy, and other evils combined. France to-day is a progressive victim doomed to extinction solely because of a failure to solve it, and if cities were not recruited from the country's virtue, civilization would be as plainly doomed as is that of France. The problem is simple: to preserve and keep progressive the human race by means of an instinct thoroughly effective for non-self-conscious beings, but ineffective for free and selfish beings who have not learned to prefer the good of all and of the future to their own selfish pleasure. In loyal monogamy the acceptance of the responsibilities and consequences of the functional instinct perpetuates the race and advances its civilization; but not only is genuine monogamy a weakening base, but loyalty to and within it, I mean the acceptance of its natural consequences, is far from the rule. The proportion of children to marriages is as steadily and uninterruptedly declining as is the proportion of marriages to population. It takes but a little general intelligence or physiologico-medical knowledge to teach the secret of the trick; the veriest booby has learned it, and Biologos is confronted with the terrible danger. Mankind must through! Self-consciousness and freedom cannot be withdrawn, selfishness cannot be transformed into love of the future race at once, and the frightful gauntlet must be run. Never was loyalty to God more needed, never was it more absent, or, in this particular, inoperative. The survival of the fit and the extinction of the unfit, even if a "true law" does not hold here, because all are hastening to become purposely unfit, and because the chief object of the process, the establishment of loyal intelligence, is thwarted in the modern divorce of loyalty and intelligence. The most intelligent refuse; the most loyal are not intelligent. The intelligent can instruct the loyal in intelligence, but the loyal cannot instruct the intelligent in loyalty. This plainly operates in extinguishing the reproduction of the most civilized, and repeoples the earth with the uncivilized. Has humanity (God) here reached an *impasse?*

The problem is complicated by the evident breaking away of this semi-automatonized mechanism from God's control during the process of humanization. There can be no possible question of our ascent into *homo sapiens* from the condition of animalhood; but the effectualizing of the advance was evidently one of those great crucial struggles of God, second only to that of gaining the very first foothold in matter. What the exact nature of that desperate struggle was, its length, its catastrophies endured or conquered,—of this we can only guess; but its scars and its dim memories yet mar our bodies and thrill our minds. One of the most striking consequences of this struggle was a change in the working of the reproductive instinct that now characterizes man from animal. Held in the leash of the divine control it had hitherto been absolutely subordinate to its purpose. Single exercise was only dreamed of for single result. But in man we suddenly find that the automatonization has broken from control, and that the implement of purpose has elevated itself into an object in itself, without restraint of season or of periodicity. Was the process of humanization so long and so desperately near failure that only by this emphasis of the instinct, this surcharging it and making it override all control and restraint—that only by this outrageous exaggeration was wreckage saved? In the animal the engine had an unfailing and excellent *governor*, that permitted just the inlet of steam required for the purpose of the machine; assuredly in man the governor has been lost, and the uncontrolled mechanism goes pounding on to self-destruction, some accidental utilization of " power " being preserved where the "shafting" has not yet broken.

But I have great faith in the divine wisdom and ingenuity. As plainly as He is often hard-pressed and near failure, as often as He may have failed temporarily and by *that* route, failure, nevertheless, is not yet upon us. The justification of all faith is present success: *We are here!* And just as certainly as we see present realization of purpose, just so plainly do we see unrealized purpose being outworked. Let us meet the future bravely and trustfully,—and helpfully. How will the depicted danger be met and the problem solved? Perhaps by these means:

1. I do not think it fancy or fanciful to entertain the hope of a final conquering of human death, in which case the whole reproductive machinery would become functionless and atrophied, presenting itself as one of the many sets of organs found in the body, remnant and outlived, the relics of ancient custom and use, preserved, perhaps, in order to call forth the conceited scorn of future " scientific " critics, who could have easily done it all so much better.* Considered carefully, there is about the mechan-

---

* But a better and more sympathetic manner of viewing these organs is coming. For example, " How reluctant Nature seems in some cases to undo her own work! How long she will allow a specialized organ, with the correlated instinct, to rest without use, ready to flash forth on the instant, bright and keen-edged, as in the ancient days of strife, ages past, before peace came to dwell on earth."—HUDSON.

ism a distinct adventitiousness of character, as if it were all a temporary makeshift, a slightness of connection with the vital organs, and a harmlessness of separation, that more than hints at temporariness, after-thought, and final laying-aside. The universal fig-leaf; the universal shame and secrecy; the silent contempt of *this* self for *that* self; the disgust of soul at sense; the commingled loathing and·yet doing—such spontaneous emotions point to the fact that God also feels that way too, and that illogic insatisfaction will some time and some way be resolved into logic satisfaction.

2. The great harassing and all-controlling task and problem of nutrition having become settled by scientific agriculture and scientific intercommunication (railroads, steamships, etc.), the constant care and labor and the frequent tragedy of the child-raising of the past will disappear, immensely over-balanced, as they will be, by the joys. This means that moderate wealth, home-comfort, and financial security, are fast coming within the reach of multitudes, and, so soon as the temporary intoxication of freedom and selfishness has passed, as it will pass, then the little solicitude and sacrifice following the acceptance of the responsibility will be gladly welcomed. It will be as nothing compared to the infinite sweetness of young life, renascent from one's own failing heart, self-revivified, and heaven influshing and overflooding the new child-wonder with the divine glory of its own incomprehensible purity and beauty.

3. The hook-and-bait idea is the appropriate and characteristic theory of the fisherman type of mind, *i. e.*, of the people who are tricking others by cunning device for purely selfish use. They that are lurers see lures. Those who always suspect others of deceit need themselves to be

sharply watched. In any partial view of the working of the function of sexualism the lure-aspect is quickly caught, but by a larger and a higher view it is as quickly resolved into larger issues and a higher unity. The salient fact consists in the recognition of abundant reward, and also in the progressive gradation of reward, in consideration of the adoption by man of service to God's purpose. It is plain enough that the beauty and charm of woman arouses love; but is not that beauty a splendid and almost sufficient reward for devotion to her, and to what her beauty means and what it stands for? With unsurpassed skill the arch pessimist has shown how every element and attribute of that beauty points to and mediates one supreme purpose and use. Is it not a proof of exquisite kindness and love that it is there, yours, an evident reward, rather than not there, and the end effected *without* such delightful gratuity, by brute force alone, as is done elsewhere, as could doubtless have been done here? Gratitude, not criticism, is certainly the fitting reception we should give both gift and Giver. "But the beauty fades so soon as the hook is in our gills." Too often that is true, and for you who do not learn the lesson that physical beauty is but a beckoning and a leading onward to spiritual beauty, the fact is sad enough. In further answer, several things might be noted. The loss of physical beauty by women is doubtless due to obviable accidents of the "struggle for existence," to obviable disease, physical, emotional, moral, and sociological, all temporarily incident to our present form of civilization. We may look forward to a speedy prolongation of the period of bloom and health so fast as hygienic life of all kinds is realized. The present woman who does not lead man to a higher delight and a better heaven than a sensual one, brings to self and husband a Pandora-box of evils. Under every other city-roof there dwell bitterness and wretchedness because one of two, or both alike, have not seen gradation of reward and progressiveness of heaven, and because they have set heart on the lower good alone. If accepted gladly and gratefully for what they are worth, the charm of sensual beauty and the deliciousness of sensual love are exquisite rewards and indemnifications. But to right minds and loyal hearts they surely and naturally lead to a better beauty and a larger love that neither fade nor falter, and that bring preferable rewards and a purer peace.

The chief of these rewards is childhood, in which heaven itself is loaned unto life. In the child the human and the divine parents clasp hands, and if the dream of abrogated death should ever be realized, the ingenuity of God will be taxed to the uttermost to devise a bewildering joy sufficiently supernal to infil, and overflood the heart haunted by the memory and the tradition, and brooding over the grief and the void of childlessness. However bitter coming death may be, we look down through tears of gratitude into angelic baby faces and softly thank God that we do not live in the deathless and childless time He may be preparing.

Next to beauty itself there is nothing so divine and of such unearthly wonderfulness of charm as mother-love. The mother and child are the inexhaustible model of artist-joy, and no true heart can, without tears, watch the lambent beauty and limpid warmth of mother-eyes, as her babe draws from the God-made breasts the milk of life. It is the very glory of God made flesh and dwelling amongst us. * Motherhood, physiological and psychical,

---

* "Of all His creatures dear to God, the most beloved is the young mother with her nursling boy. Of all earthly loves, the most holy is that of this mother for her child. Of all human suffering, the most poignant is that of the young mother struck by death and leaving her darling to a probably tragic fate.

"This love you received, this love you gave, this grief you endured,—for

can tell more about biological philosophy than all the books that have been written on " Evolution." Everywhere we go the vision is before us,—for what animal does not love its young ? There is the same angelic benignancy in the eyes of all mothers, animal or human, because both mothers are filled and thrilled with the same divinity. One God is in both hearts outworking one purpose by the same emotion. And what grief so deep as denied motherhood ? What sight so sweet as the stifled hidden love of childlessness finding opportunities for its outgoing ? A child's doll is a holy thing.

It is not only the mammæ of the childless woman that swell with milk at the cry of another's infant. I know of an instance in which a little dog made a moving family most miserable by its whining, moaning, and scratching, all to reach a mewing kitten kept in a basket. After several days of this torture dog and kitten came together. The little doggie had had no puppies for several years, but the kitten's mewing had aroused the glands to secretion, and, in nursing the motherless kitten, the dog's satisfaction was surely as great as was that of the foster-kitten. I have seen an unsatisfied hen drive a cat away from her kittens, guard, and watch them, fold them under her wings at night until they were grown. Hudson tells a story of the splendid heaven-exploring chakar. One had been domesticated and made a household pet. It loved young chickens. A few were given him, and so perfect was his care of them that many were allowed him. " It was very curious to see

two years,—and of these years each bitter day was an agonized prayer that your life might become my life, and your Father my Father.

"Dear dead, undying Mother, I would I had something better than this poor book to offer as a sadly belated *Amen !* to your prayer."

[From " A DEDICATION."]

this big bird with thirty or forty little animated balls of yellow cotton following him about, while he moved majestically along, setting down his feet with the greatest care not to tread on them, and swelling himself up with jealous anger at the approach of a cat or a dog."

4. Future sociology and government must undertake a certain ordering and regulation of the reproductive function. God is waiting to turn the task over to man. He has brought man to self-consciousness and control, and the sway of His instincts will hold the race in being until the deputization of His power shall have been effected. Marriage and procreation must, by stringent legislation, be absolutely interdicted to those without physical and moral health. This necessary step should be at once resolved upon and carried out by all civilized governments. There is much to be said for the proposal to deprive criminals of the procreative power. It is, perhaps, good sociology and good penology. The taxation of bachelors; the control of the "social evil" by wise and careful legislation; the severe punishment, not of the poor mother alone, or chiefly, for criminal abortion,—a fact, alas! more common than is guessed; the making disgraceful the "legalized prostitution" of intentionally childless marriage,—these or other better methods of controlling private selfishness for the common and future good will be undertaken, and will help to work out the salvation sought by Biologos, and which is directly denied Him, because of the enormous difficulties of His means and of His obstacles. Here as everywhere God is calling, waiting, begging, for the sympathetic help of loyal men.

5. Humanity is now in the beginning and press of an advance such as is comparable in importance only to its

spring out of animalism. The great unrevealed mystery and glory we dimly prefigure and vaguely surmise under the name of civilization, is pre-eminently an advance of reason and of the common well-being.

General enlightenment and comfort stand out as distinguishing occident from orient, and present from past. The most untrue rant of anarchist, or wail of pessimist, is itself proof of progress attained and unattained. But all this is only the preparation of the ground. I prophesy that the "religion of the future" will be summed up in the words intelligent loyalty; loyalty, individual and collective to God, and to the purposes of God, as revealed in the scientific study of biology and history. Whether this faith of the future will grow out of, or attach itself to the historic continuity of Christianity, no one can say. If Christian leaders were wise, the historic momentum of that great and sublime faith could be utilized. The ground is all prepared. Christ's teaching lends itself easily to this utilization and application. But the plasticity of old institutions is usually sadly deficient. The Church should slough its trinitarianism, its credal and aberrant characteristics, and grasp the opportunity that for the last time is offered it. Christianity has been the enemy of reason, of science, and of progress. Her spirit has not been the promoting but the hindering spirit of civilization. It is useless to say that this was not genuine, that it was not the Christianity of Christ that accepted asceticism, martyrized Galileo, and, if Luther had not been a Teuton, would have quenched him with the ever-ready fires of the Inquisition. The argument has no force. Let it be admitted or denied. The question returns, what of to-day and of the future? With not the faintest spark of antagonism, with perfect absence of prejudice for or against, let one ask if on the part of present-day official Christianity there are any signs of intellectual grasp of, and Christlike devotion to, the real problems of civilization and the future? The Christ of Nazareth naturally and excusably knew nothing, betrayed no slightest hint of knowing the uses of government in bringing about the Kingdom of God on Earth, of the scientific study of sociology and penology, of the incarnation of God and revelation of Himself in the biological world, of the power of education and of printing, of the glory of Science, of the means of commerce, and especially of modern mechanics and intercommunication, to help on the realization of his ideal. But the Christ of to-day must know and grasp these opportunities and work in and through them. The simple question of the existence of institutional Christianity lies in her possibility of plasticity to adapt, and adopt, and in her vigor to seize upon these things. Whether she do this or not, will not deflect humanity from its evident onward march, and other institutional forms and spirit will take up the work. There is no escape from religion; the joy of human existence will depend upon its vivid reality, the motive of true enlightened hearts will consist in loyalty to God, the living God that is working under infinite difficulties in and all about us. This intelligent and active loyalty to an intelligent and working Father will catch life, and fill heart after heart until it will become the renovating thrilling spirit of coming humanity. With this enthusiastic loyalty the dangers from the exaggeration and lawlessness of the great instinct upon which our existence depends will be slowly solved, order and control reinstituted, and the great work of the overcoming of death itself be undertaken.

CHAPTER VIII.

CONCERNING EVIL.

THE origin of evil has been the insolvable problem of all theologians, and there is little wonder that it is so. There is absolutely no escape from the terrible logical dilemma in which they found themselves placed. One of three things must be acknowledged by any mind not the most logically depraved: either God is incapable of destroying evil, or He does not wish to do so, or no evil exists. "The grinning atheist could not be answered"—from the "good father's" standpoint at least. To the latter it seemed the worst sacrilege to question God's capability or omnipotence. Hence all theological discussion has centered about this insoluble question, concerning which there has been even more hair-splitting and wrangling, more intellectual bravery and cowardice, than over that other sorry problem, the existence of God. The faithful clung to their omnipotent and infinitely benevolent God, as they should have done, and tried to minimize and forget the frightful fact of evil. But it can hardly be doubted that there has always been a large minority of virile intelligences that were driven into atheism rather than deny this appalling verity of evil. Busied with their thought-spinning, all forgot the simplest duty and easiest way out of the intolerable mystery. All logic-choppers are everlastingly hunting for their spectacles everywhere except on top of their head. One minute's observation of the living world about them would have made clear the plainest of facts: the fact that every organism is struggling against difficulties to maintain its own existence and to bring others like itself into existence. There is no mistaking the meaning of these facts: God is working under difficulties imposed upon Him. But to the poor theologian this would have been impossible to see or to acknowledge. It would have been too scientific an observation to make, *i. e.*, it would require perception of fact, a feat quite impossible to him, and it would have destroyed his ideally conceived, far-away infinite God. Thankful are we that our minds and eyes have been unbound and are opened; that facts and their meaning are our study, and that the living, acting God is the object of our love. Dreams and word-weaving, mental tyranny imposed or chosen, dead gods and law-bound life,—all are banished, and with eager clear brain, fearless, and yet reverent, we walk in freedom and among realities.

As to God's goodness, I would be perfectly willing to see or acknowledge any proof or intimation of moral imperfection. I fear no truth, but love it above all things. Neither critic nor apologist of God would I be, but a knower. And still I say that a keen eye penetrates no mystery, and that a balanced judgment ponders none, that does not show benevolent love at work. It is perfectly possible, for example, that justice should be stronger in Him than love, but I see little evidence in life or in history whether biological or human, that God has much, or indeed any, of what we call justice. He is not unjust of course, if He were He would not be kind and good, as He is; but every indication of His character that I can see shows beauty and love conquering justice, dragging him to play as laughing children do a make-believe-serious dog or father. The Greek Olympus was an infinitely better and truer idea of heaven than that of any Christian painting I have heard of. The laughter, the fun, the song, the play, and the spontaneous joy that breaks out above the struggle for existence whenever this bitter struggle may for but an instant be evaded or forgotten, the children, the birds, the flooding of utility with the sunshine of beauty,—all speak of a God who Himself laughs, loves, and ornaments, and to whom the morbid imaginings of medieval hells and puritanical heavens are alike repulsive.

There is no honest eye and loyal heart that can blink the tragedy of death. To lose the suppleness and sense of elation of strength; to feel the labor of energy coming on and growing greater; to be compelled to watch and guard against our little personal weaknesses and dangers; to feel the waning not only of physical zest, but of mental and emotional *verve* and *élan* ("he begins to die who quits his desires"); to know that the end is coming, yonder, soon maybe, a little later certainly,—all this admits of little mitigation,—we shudder and brush away the tear and the thought. Well! I frankly say the shudder is God's shudder. It is sad and tragical, and the tragedy of it crept out of God's heart into yours. But it is not God's fault, and He is doing all he can to circumvent, to obviate, and to overcome it. To the circumvention of death He has superadded a royal bounty of beauty and blessing in the loveliness of love, of flowers, of maternal love, of woman, and of all that rightfully pertains to reproduction. Gratuity is heaped upon gift until the eyes are dazzled and the heart sated,—so kindly thoughtful has He been, and so solicitous to make us forget the ugliness of His enemy and of ours, and to recompense us for the pain we are put to in dying and in passing our life on to the child. Death is almost unmixed evil, the price paid to matter for a temporary service. Its scientific aspect is simply that Biologos has been unable to maintain permanency of nutrition of the individual organism. He has attained it for the type by the device of reproduction. There is hope that it may be yet attained for the individual.

The same difficulty of nutrition explains a large part of the evil of disease. Every -pathy is at bottom a trophopathy. Much disease is due to man's physical sin, much more to his moral sin, and still much more to his intellectual sin of improvidence. But there still remains a large proportion that must be charged to the inherent difficulty of the incarnating labor, *i.e.*, it is fundamentally nutritional, and for long will remain unavoidable. Disease and subnormal length of life are appallingly evident in the human being. Such is the expense, the difficulty of humanization.

There is one aspect of pathogeny that needs to be lighted up by some philosophical mind: diseases are often hyperphysiological in absolute etiology, and their prophylaxis will be found to consist in the laws and rules of the best psychic life. I mean that the obviation of disease and the progress of physical health are attained only by means that obviate moral disease and insure spiritual health. Take the entire class of crowd-diseases. Too great crowding also produces psychopathy, and hinders mental stability and progress. We must be much alone for the best health, either of soul or of body. Extend the term to include most contagious diseases, typhoid, cholera, diphtheria, etc., because most such diseases are crowd-diseases in a strict sense of the word. The open sky, not roofs, above; the pure ground, not pavements, beneath; the free space, not poisoned air, about us, are as necessary to the best spiritual as to the best physical health. The intimate relation of all contagious diseases to uncleanliness and of wound-healing to cleanliness, also points the same lesson. There is a close, though not a rigid, and a progressive relation, of cleanliness of body and cleanliness of mind. In luetic disease the connection of moral and physical evil is most true and most awful to behold, and allows the suggestion to arise that here almost alone in the entire range of biological phenomena does one meet with a fact that suggests vindictiveness, or at least a punitive character in God. As in regeneration one sees the astounding evidence of His hand, in boldest miracle of action, so in venereal disease, one seems to see the clenched fist. Only the physician sees and knows what may not be described, but to his keen and kindly eyes a thousand mysteries are revealed, and a thousand tragedies are justified, that stun the mind and heart of the ordinary citizen. In pulmonary disease, tuberculous or not, the ultimate causation-agencies are occupations, habits, and conditions that kill the lungs of the soul and of the body. There is a multitude of conclusive evidence that the bacillus tuberculosis and most other pathogenetic bacteria are powerless against properly developed and normally vitalized tissues. If given proper blood and air and exercise, the lungs offer no nidus or hostelry to the little enemies, and proper blood, air, and exercise also bring a power of mental and moral respiration that resists the horde of psychic bacilli hovering to destroy the spirit as well. If this is all true, then the healthy, full-breathing, pure-air-loving lung is the standard, and the bacillus is the agent that punishes the sin of falling below that ideal. If this agent were not operative, the narrow-chested, city-minded, deoxygenated degenerates would be the parents of a progressively degenerating race, and a successful Koch its worthy high-priest. Fortunately the "lymph" did not fulfil expectation. How far the great historic epidemics and scourges of the past and the diseases of the present have been the result of moral sin, is beyond our power to discover. That the brutal selfishness of rulers and aristocracies has been a most potent agency in producing famines and wretchedness, the prolific parents of disease, is too plainly evident to question, and of which the France of the past and the Russia of to-day are lurid examples. Then, too, there are a number of diseases that in a special sense may be called civilization diseases, scleroses of many kinds, cardiac and renal inabilities to endure the strain put upon these organs, neurasthenias, psychoses, hysterias, hypertrophies, hyperplasias due to alcohol, excitement, physiological misuse and disuse,—all flowing from or intercurrent with ethical misuse and disuse. The most incongruous absurdity imaginable is a materialist-physician. But the inquiry is too vast and as yet too vague, for more than mere allusion to its possibility. In a world in which the entire biological process, every inch of ground gained or to be gained, is gotten by infinite watchfulness and ingenuity, ethics must necessarily be largely hygiene, because every physiological question is ethical, and every ethical question is physiological. God Himself being the great physiologist, His power and means of incarnation being physiological, moral evil and pathology quickly reach approximate identity.

All of the evils attendant and consequent upon the survival of the fittest and the struggle for existence are easily recognized as subordinate to the nutritional difficulty of Biologos in getting and in retaining foothold, or to the ideal of progress that all sacrifices mediate. The bellum-omnium-contra-omnes, and the nature-red-with-tooth-and-claw philosophy, as all now recognize, has been greatly over-emphasized. There is no doubt plenty of fact-basis for the theory, but when seen in large and calmly, it is all presently necessary to preservation of balances, and brings about perfection of organism together with progress of type. All things considered, the death of the supernumerary, and of the less fit of the animal and vegetable world, when it assures the realization of right purpose, is itself right. The pessimistic finding that the process does not lead to survival of the fit and to ultimate progress, is, like many things, "critic and whipper-snapper," simply untrue.

"The slow movement of history, the stagnant nations, the dead levels, the non-progressiveness, or even back-slidings," and all that! Readily acknowledged! But have we realized His difficulties? Have we even acknowledged the existence of them till now? Are we wiser or more powerful or more ingenious than He? Would He not have hastened if He could? Is it not He that has brought the present success about, even so late as it is? Are not we the results? Nay, is it not in reality early, and has He not truly hastened? When so much that is good is seen, may we not agree that the bad and unseen has been as good as it could have been? Not His the responsibility, but the unconquerable irresponsiveness of His material. No blame to Him, but infinite gratitude for a thousand huge difficulties overcome, and for success now and yet to be wrung from Fate and Death.

"But men are wicked and life is embittered by their hideous selfishness." Sometimes, and too often true, but not all men, and not any man irremediably or wholly. How often, if we will to see, do we not learn how good even our worst enemies and the worst of men are, and how true it is that only opportunity is lacking, the spring-warmth of love and kindness from us, for example, to make goodness like a cactus-bloom burst from the very heart of thorns and ugliness. Do not believe men are wholly bad, or that they love evil. There is doubt in our minds as to what things are evil, and what good; conditions unhealthy to body and soul have warped and stunted and deranged the mechanism, but such conditions are in process of passing, and man's spirit, a graft of the divine life, turns loyally to the divine ideal, and turns just so fast, and just so far as it may.

People flatter themselves with the illusory greatness of their adversity, their martyrdom at the hands of niggard fate or of cruel men. This heightens their little heroism, and makes them seem greater to themselves. We are all quite expert at playing the courtier to our own ego. But after all we well know in our hearts that most of the really harmful evils of our lives are the products of our own folly and wrong-doing. Another can harm us very little unless indeed we kindly aid him. A little careful introspection shows us that very often we love the very faults in ourselves whence flow the conditions at which we grumble. We do not desire to be better, and would be sorry if we were forced to better ourselves. Perhaps there is a subtle wisdom in this that smothers our smile. Every character has its own laws of balance, and its peculiar difficulties of function, known only to the silent wisdom of "the unconscious," and our little obstinacies, frictions, and malfunctions are often devices for preventing greater evils and of keeping well on the track.

Then how well we know that, if we could rob them of

it, the wealth or the power that we envy in others would be our own undoing. How few people there are that are not injured by their prosperity. They may very rarely use their money or their position to help others, but by positive wrong or by negative non-doing, to self-injury they add injury of others. For the present degree of elevation of personal character there is entirely too much exceptional and personal " prosperity." No greater kindness could one wish most of his friends than misfortune. The crude, hard, cruel plebeianism of " successful " people; the waste of their leisure; their emulation of people more disgusting than themselves; their greedy hunger for more; their waste of what they have; the degradation of their petty imitators; the hiding of their interior poverty under the hollow show of luxury, and of their irreligiousness under the impious hypocrisies of religion,—all such things lead to the conclusion that without accepted accountability deputed power ends in evil done and evil suffered. Riches without character to use them wisely either for self or for others is the consummation of evil, and that government or that society that produces this condition so easily as do ours is travelling a dangerous road.

Not only is wealth harmful to most people, but poverty is a blessing to most. The majority, as regards the usefulness and excellence of their character, have all the prosperity "they can stand." If the poor could forget envy, and fill their hearts with joy, how delightful their lot! Happiness, as any darkey or dog might show us, has little to do with circumstance, but is instead a question of temperament; the most miserable of people, as we all know, are the well-to-do. The " poverty of wealth," is the most pitiable of all poverties. But there is nothing like want and struggle to make us sympathetic and heedful of others. It is only through suffering that we see the mysteries of life, and the goodness of our fellows, of the world, and of God. There are few people that would not be the better for a severe illness every few years. Of two sisters, once alike in youth and health, how inexpressibly better and nobler the one that has passed through trial and suffering. Of two boys, one the product of the city and of comfort, the other, of the country and of aspiring struggle, how infinitely superior the latter. Want and the overcoming of want have given him eyes to see, a heart to feel, and a hand to do, and upon these things the destinies of the coming generation depend. First of all other attributes of God must be placed that of activity. He is a worker and therefore is He happy. The *ennui* of the divinity of the old theologians must have been as insupportable and as pernicious as that of a modern professional human do-nothing. But the uses of adversity have been the subject of homilies since preaching was, although its correlate, the utility of much that is called evil, is hardly acknowledged, and less still, its necessity.

The evils of the world are therefore divisible into two great groups:

1. The unavoidable, or those inherent in the difficulties of the incarnation process. Death, at least for the present, must come to all; many diseases belong here; the necessary exaggerations of automatism, and the hypertrophy of functions; those unavoidably springing from the agencies mediating progress, such as the animal struggle for existence, the nutritional difficulties in a world of physics, etc. Then, behind all, is the control of fate and of accident. The whole incarnation-process, at present at least, is an adaptation to and into a world whose large

forces it cannot touch or control. The secret of that process lies in the use of the infinitesimal forces of nature; only by the astonishing indirections of cytology and of animal life can Biologos reach a slight degree of control of the larger or molar forces, and only through man can He increase a little that power, and make its use serve purpose. Thus the foundations of our earthly house of life rest upon the shoulders of the giant forces of the inorganic world. Every quiver from below sends a shudder through all living things. Neither man nor God has any slightest control of the size, the qualities, or the supplies of atoms; any control of the sun's heat, or of his forces of attraction, electricity, and magnetism; of the composition of the earth's crust; of such conditions as earthquakes, tides, rainfall, or cold; of the constituents of the atmosphere, and so on, upon which are based the infinitely ingenious adaptations and difficult maintenance of all living things. When the blows and tragedies of these fates fall upon us we may uselessly blame dead forces, but the truer attitude is to pick up the work again, help each other, and especially rally loyally about our Great Helper, who Himself suffers with us, and who is more able than we to build victories out of the appalling routs and the most chaotic ruins.

2. Finally we have the evils that are subordinate to the mediate stages of the process. These are mostly temporary and in time will be avoided. In the interim these arouse our energies of life, condition our progress, and are educative. Such evils are the wastage and wreckage of young life; the less fierce aspects of the struggle for existence; the preventable diseases; the unequal and unjust distribution of property; the hypertrophy of egotism and individualization (a consequence of the furious

struggle into humanization); the tyranny of rulers, and many such. The unrolling of the panorama of earthly life is a process; in a rigid sense of the word all imperfection is evil; but any motived process implies temporary imperfection and direction somewhither. Moreover, the possession of freedom implies some misuse of that power, otherwise we should be the merest tools and ruled by external forces alone. To learn right use, presupposes some abuse. We learn to stand by having fallen many times. Experimentation and experience imply failure, but they are the conditions of functional freedom.

CHAPTER IX.

JUSTIFICATION OF THE INCARNATION PROCESS.

FOLLOWING close upon the question of evil comes the pessimist's conclusion that life taken in its entirety is evil, and that no life is preferable to any life whatever. This, of course, is unscientific, because it is an abuse of induction. It posits a universal conclusion upon a very limited gathering of facts. Even if all life in the past and that the pessimist could know in the present were not preferable to nonentity, it would not prove that all coming lives would be thus condemnable. But the pessimist's syllogism is an outrageous *non sequitur*. We can excuse Buddha for assenting to the logic, but the modern echoist is contemptible. We all know that with the vast majority of people there is more happiness than unhappiness. One proof is that they continue to live. Though Schopenhauer wrote *Nichts* over the future, and said *Nichts* was preferable to *Wille*, he nevertheless stuck to *Wille*, and the whole machinery of Nirvana and the struggle to reach it has little more *raison d'être* than an excuse for continuance of what the creed said was unworthy of continuance. Every suicide who clinches the pessimistic argument with the evidence of sincerity, we also see, acts from poor reasons. He fails to perceive facts in their proper relations, and to adapt his resolves so that they shall be in harmony with them. Suicide is nothing

JUSTIFICATION OF THE INCARNATION PROCESS. 189

but folly, a child's pouting, and *I won't play*, because one cannot have things his own selfish way.

The question however, arouses the further inquiry as to the ultimate reason for the whole process of incarnation. Few people see any object in their own life, or in the life-process of the world. The beliefs in the vacuous heaven and the do-nothing God exaggerate the tendency, and some day, like Doré's neophyte, the startled mind flashes a wild glance of inquiry about itself, while cosmic fear congeals the blood with the panic of discovered universal objectlessness. It makes the heart stand still with curdling horror to feel that blind forces and processes rule all, and that our little lives are driven eddies of the dust of chance in the gust of circumstance. Frightened by the thought, countless numbers of minds have shrunk trembling back into the consolations of an authoritative religion, or dashed with brute bravado into the equally tragic death of atheism and pessimism. If I should be the means of showing one "neophyte" a third and infinitely truer course, I would consider the sacrifices made to bring this little book to his notice, and even those of my life itself, as worthily rewarded. I would like to spare my unknown brother the suffering of gaining the lesson in the way that I have gained it.

The bounteousness and benevolence of the Father of the living world is shown in the fact that whilst each form or type of form is more or less dependent upon other living things, each is so filled and lighted with satisfaction that it is "its own excuse for being." It satisfies not only the self, but it also satisfies the on-looker, even if there were no "beyond." This is especially true of the vegetable world. Whoever has stood in the center of a landscape devoid of all animal and plant-life will always carry with him a realizing sense of the homelessness, horror, and void that our world would show without a living green thing on its surface. Our own satisfaction and delight in the inexhaustible loveliness of grass and grain and tree cannot lessen while the planet is our home. But we must not forget that it is also God's satisfaction and delight, and that it was His before it was ours, and that it is so to us because it was and is so to Him. He taught us all we know of agriculture, of gardening, and of forestry. As we look through a powerful telescope at the dead moon we justify God's work of the plant-world by our shudder at the desolation there, and by our wish that it might be as rich with grass and tree as our delightful home. If we may not penetrate the divine mind or discover His motive as we may suppose Him prepairing for the beginnings of living forms here, we may rest content with the childlike thought that it is infinitely better that our world should be decked and beautified with green life as it is, rather than be as desolate as the moon is.

For my part I would be quite contented to let the idea of final cause rest, happy to know God's "most ultimate" motive may have been simply the play of spirit, and that creative activity is the natural inherent attribute of his being. But every act of His looks to a beyond. Any idea that our poor minds can shape as to "that far-off divine event to which the whole creation moves," must fall short of what that event will prove to be. The fact of process, of the whole proceeding somewhither, is amazingly evident. Some purpose is gloriously clear, but just what it is we can only cloudily gather, and we know that in such far visions all but the largest outlines must be dim. We dare not be dogmatic; even in our little scientific and mechanical discoveries we do not at first see what will be their best and final uses. When we observe the mechanism and functions of a grass-blade, of insect and of animal, of man and the established harmony of the world of these things, we may be more than certain that the Mind that could do all this, and the unseen-more of the future, is incapable of working aimlessly, and that purpose runs through all and will so run to the end of time.

The essential secret of incarnation was discovered and the fundamental problem solved with the plant. With that was proved the ability to build living forms out of dead atoms and mechanic forces. But onward presses the hidden design! The walking plant, the animal, is evolved. By the secret of appropriating the products of plant-life, the problem of the nutrition of the moving mechanism is solved, and the higher sensations are born, seeing and hearing being tools and aids of motility. Ever onward! Out of, and by further use of the animal product comes man, with his new endowments of mental vision, and his beginning purposive control of all mechanic forces, but especially of molar motion. Civilization is first a fixing and stabilizing of the certainty and uniformity of nutrition, but it is already largely being turned to ends beyond those of mere utility. Pure science, higher knowledge, discovery and the control of various natural agencies, are seemingly preparing the way for a progressive mastery over the entire world of mechanic forces. Every discovery lured into being perhaps by hope of reward or of its usefulness to man, seems destined to pass beyond any simply utilitarian aspect, and to be gathered to a higher aim. For example, every discovery in mechanics was needed to perfect the best astronomical instruments. But every discovery points to increasing dominion by man over the physical universe. Interplanetary communication is not absurd, and even more than that. Every study of life on the globe shows progressive power to build up living forms out of dead matter. More and ever more of the atoms of the physical world are being caught up and pressed into the control and service of Life. All below man attests progressive control of molecular and atomic forces, and through this and through man's science, there is dazzling proof of coming control over masses of matter and oceans of imponderable forces.

What is the dim but ever-clearing suggestion and indication? The vivification of a dead universe and the purposive use of mechanic forces. Uselessness is to be changed to usefulness, objectlessness to rational ends, meaninglessness to significance, death to life. In His infinite love and activity God seems determined to transform even the dead stones and dead worlds of a dead universe to the living image of Himself. Without life and its attendant purposiveness the physical universe has utterly no use or meaning. It has always existed, perhaps with the great pulse-waves of eternity—long integrations and disintegrations described by Spencer, but all the mere empty play of mechanic forces, and, without life, leading absolutely nowhither. With the intervention of Biologos, however, entirely new elements are added, all is suddenly changed and the world is palpitant with significance, interest, and purpose. Physical forces are reined and ruled by intelligent design. The control of matter and of its aimless forces is the summarization of the primary function of both God and man. If by God, then of course by man, because man may be defined as first the tool, then the deputy, and finally the co-worker

JUSTIFICATION OF THE INCARNATION PROCESS. 193

of God,—steps in the extension of the control of matter by spirit. It is noteworthy that the conscious discoveries and inventions of man already made are but the beginnings of imitations of God's inventions; we are but rediscovering new continents He had already surveyed. He found Himself unable to clear and till the soil by direction, and hence He is indirectly doing it by our hands. It is all at His suggestion. Science is His work of colonizing and civilizing His own country. There is not a single mechanic discovery or art of civilization that has not been used by Him first. Every application or principle that we use in bridge-building, architecture, telegraphy, motility in air, water, or on earth; every use of chemistry, heat, light, electricity, or magnetism; every device of assault or protection; every tool or instrument whatsoever that we have made—all have their analogues in animal or vegetable life; all were His before they became ours. Our works indeed are not seldom mere adaptations of His mechanisms, and the fire of science is bright because we are only commencing to learn. If we only had the knowledge and power over atoms and molecules, *i. e.*, if we were only such biochemists as He! Could we but fly as His birds! We have not yet the faintest idea of the mechanism of muscle-contraction. If even we but knew the mechanisms of sensation! If our artists could but mould and color as He can! What musical instrument can compare with the human voice! Our poetry is the attempt to re-sing His songs; our tragedies are the poor echoes of His griefs; our joys the reflection of light from His heaven!

We can do nothing, could not live a second without levying upon and often most ruthlessly using the inventions of His hands. Think of what our lives would be without the wood of His trees, used in every thing necessary to our comfort and to our very existence. Even the minerals would not be attainable or manufacturable without His help in wood, fuel, and mechanisms. There is not an article of clothing that is not the appropriation of His work, the most useful and serviceable being those that have most occupied His care, and that have previously been of most use to His animals—wool, hides, and furs. If we come closer to our life and think of our food, we are quite ashamed because the brutal selfishness of our greed becomes painful, All our meats are the muscles of His beloved animals that we have killed; all our fruits and grains and vegetables are the stored supplies of nourishment for His seeds; eggs are the material and food provided for making and nourishing the feathered young; and milk, the purest, best, and loveliest of foods, has learned the secret and beauty of life in the beating hearts of the purest, best, and loveliest of His children. Surely our indebtedness to Him not only for our original being and the upholding of it by the marvels of cell-life and nutrition, but also for every instant's supply of tool and mechanism, clothing, and food,—truly these facts leave no excuse for the doubting of our primal duty, or for not recognizing His work in ours and ours in His.

From the lowest unicellular organism to the highest human scientist and discoverer, all living function consists in progressive control and intellectual use of physical forces, and this is the characteristic promise of the future. Already we are looking out into space to extend and to see evidences of the extension of the same control and use. If we judge correctly, but few of the planetary bodies of our system permit Biologos to carry on His work on them. In the sidereal heavens there are probably but comparatively few worlds whereon He has been able to gain a foothold. What a link our cosmic life may be in the whole plan and progress of the universal life-system we cannot guess. We may be the outermost colony of some swirl and cluster of sun-systems, and from us it may be planned to transplant to new solar systems beyond and ever beyond, the mechanisms of cellular or organic life without the preliminary stages of our own past struggle for existence. The imagination may play with the beautiful thoughts aroused, and strain eye and brain to catch glimpses of the inter-related harmonies and purposive adjustments of the future extension and progress of life, but as to-day's realizations in civilization are infinitely richer than the most superb forecastings of any medieval fancy, just so, we are convinced, will the future transcend the most dazzling speculations presumably possible. The essential thing is hope and confidence and knowledge that we are parts of a great system of purposive progress; that our duty is loyalty to the evident design and Designer; that the fundamental mechanism on which all depends is spirit-control of physical forces; that the condition and motive of all progress is extension of that control; that the essence of all humanization and ethics is spiritual use of these forces.

There is the plainest evidence of a progressive conquering of the nutritional difficulty in the future biological process, and with this will come the double victory of Biologos over the dead and stubborn world of crude matter, and over time itself. Only an infinitely small part of the universe at present shows spirit in control of matter, but the prospect opening to our view is that of a progressive dominion of God in the universe until the loneliness and death and uselessness of all the lifeless worlds of space shall be flooded and filled with the joy of intelligent life, and of that life progressively perfecting and extending itself.

Something such, it appears to me, at least temporarily and tentatively, is a sufficient justification of the incarnation-process, and a sufficient atonement for the labor and evil of its past stages as read by science, of its present stages wherein we are the actors and observers, and of whatever the future may bring. I find in this way of thought a motive for all energy, an explanation of the great mysteries, an ideal of all conduct and aspiration, a principle and actualization of home-making in the universe, that brings a blessed peace to my mind and heart that is in most beautiful contrast to the tragic failures of past philosophies and religions.

Wherever we look we see mentality and spirituality extending their control over matter and shining through organization with eyes ablaze with desire. To illustrate how purely psychical or metaphysical qualities dominate those that are purely physical, and thus give the victory to the larger and better type of character, would be a a work of supererogation. In every day of our life, too, the lesson comes home to us that in the long run the quality and motive of mind outvalue the degree and power of mind. While the young Napoleon's mind was plastic and patriotic, nothing could conquer him. Later, although with more infallible judgment, with ripened experience, with systematized powers and every prop of government and objective force to aid him, he failed. He ran headlong upon the world's moral sentiment, incarnate in Wellington and his English boys. Why? Because honor and a noble ambition had given way to selfishness and ignoble ambition. The lower had displaced the higher, and the physical ruled the spiritual.

Examples are all about us: Even in such a purely physical encounter as that of the prize-ring, Corbett whips Sullivan; which means that the better mind and heart (character) more than make up for lack of pure muscle and physical powers. Courage is etymologically a matter of the heart, and of the psychical heart. The body follows loyally where the mind commands strongly.

Paderewsky has inferior physical, even inferior (technical) musical ability to other players, but his playing is soul made into sound, and, expression being all, the large pure beautiful character of the man inthrills and overfills tone with the heaven incarnate of divinely psychic revelations.

Every noble physician knows that it is not absolutely the most scientific man that succeeds best in luring the departing soul back to the sick body, and holding it to the work of life. Where one with all the therapeutic erudition of the world will fail, another, a simpler but holier physician, will enter the patient's chamber, and lo! on the walls instead of the sentence of death there gleams the legend of convalescence. The heart and morality often reach intellectual vision that is denied to the strongest intellect.

In politics, even, there are occasional promises and examples of the same law, the most striking of which is the late peaceful revolution that the whole American people have effected and seen. "Bossism" and the degradation of principle have found that there is a political value to probity and psychic valor.

It is only into the vilest walks of life, in the realm of ignoble money-grabbing, that the law has hardly found an entrance. Here one cannot say much more than the prayer, How long, O Lord, how long! Wherever else we turn, in art, science, literature, or human intercourse, it is the quality of the psychic life that lends biological things their value, and that endows them with charm and durability.

Let us also not lose sight of the everlasting fact that this peculiarity or quality of psychic life is moulded or changed, like light passing through the sun's photosphere, through the telluric atmosphere, or through a colored lens. The pure light and the pure divinity are by the traversed media changed only by subtraction of valuable elements. Organization (sarcogenesis) is the lens or turbid medium that allows the divine light of the life to shine through it more or less with difficulty, subtraction, and distortion. But the life and the light of heaven are always seeking to reach us in all their pristine purity and strength. The perfect physiological and neurological organization is that that best transmits the life untouched and unclouded, even, if possible, focusing it in a purer and warmer beam. The problem of morality, of art, of heroism, and of saintliness is the problem of achromatism. Let us pass God's life through the atmosphere of our earthly life and the lens of incarnation untouched and pure, and in character let it be focused in the glory of the mind that is at once both divine and human.

CHAPTER X.

FREEDOM.

AFTER the problem of evil, that of the freedom of the will has been the next most puzzling difficulty of theology and philosophy. It is, indeed, only another phase of the first. On the supposition of an infinitely powerful and good God it was impossible to solve the question of the origin of evil: either evil did not exist, or else language and thought were useless. Just so as regards freedom: if omnipotence and omniscience created the world, it is of course impossible for human freedom to exist. Dishonest and cowardly intellects may juggle with the question, and the record of past jugglings is either pathetic, or ludicrous, or both, as one may look at them. The splendid acrobatic performance of Calvin and his followers is the crowning act of the tragical farce or farcical tragedy. By dint of a frightful solemnity of manner and countenance, and with a hypnotic fixation of the attention upon a lot of words majestically intoned,— foreordination, predestination, and all the rest of the pompous Latin polysyllables—one comes near forgetting logical facts for illogical fancies.

Hypertrophic or parasitic intellects, that is, those developed at the expense of hypotrophic or atrophic emotion and perception, have always been deterministic in belief. Such minds have no difficulty in disregarding the ineradicable human consciousness of freedom, and the plain objective facts of life. They always prefer a subjective process of thought to a recognition of objective facts. Whether the logic-chopping machine be worked in the supposed interests of religion, or of philosophy, or of science, matters not. Calvin and Schopenhauer and Spencer strike hands on this point. And when the logic-machine butts against the immovable wall of fact, it turns tail, and points the other way—*i. e.*, facts are ignored, the belief continued, but much weakened and made foolish by a useless persistence of wheel-turning without perceptible movement forward. Now according to my way of thinking, God, evil, freedom, etc., are not alone or chiefly matters of subjective thought processess, but if they exist at all are genuine facts, to be perceived, studied, and tested, exactly as one studies a chemical reaction or a physiological process. They are true objects of scientific investigation, and are to be submitted to the discriminating reason and perception with questions of how much, how far, what qualities, and what degrees, precisely as we approach any biological investigation. If trinitarianism be a fact, it is a matter of scientific investigation, as are all facts, and the articles of the Apostles' creed, or of the Nicene creed, are to be considered as subject to the provings of historical and biological investigation, and in exactly the same way, as Weissmann's theory of heredity, or the influence of Occidental civilization on Japanese development. And the *onus probandi* lies, of course, with the assertor. If one assert that Japan developed its present governmental and social changes because of the influence of Aztec immigration and discoveries, it behooves him to prove it by historical facts; if the theory is an illogical dream of the fancy or of superstition, let it so appear. Just so with any one of the Thirty-nine Articles.

It appears to me plain that every step, nay, every creature, of the biological process shows adaptation to and utilization of "the environment." Circumstances by no means dictate the "reaction," but the organism, whether plant, ameba, horse, or man, "reacts" to "external stimulus or environment" in just that way that subserves the object it has in view. The reaction is never purely mechanical, but is always hypermechanical, always one of utilization, of selection, and of purpose. Now, the entire being that uses, selects, and designs, is something that comes into matter with the evident attribute of spontaneity, choice, or freedom. Precisely as an architect choses this or that plan, this or that material, just as he changes, shapes, supports, ornaments, or colors—so does Biologos, in coming into His circumstance, and in using His material. And precisely as the architect makes his building conform to the purpose and needs of the occupant, so does He; and precisely as the architect is forced to conform plan, extent, character, etc., to the materials procurable, and to the means at hand, so does He; and precisely as the architect whenever he is able, makes the appearance of the dwelling to others and to the owner, beautiful and delight-giving, so does God in all the works of His building. But the architect has freedom, as houses good, bad, beautiful, and hideous convincingly demonstrate. That God's creations are always most perfect and beautiful, always so much so as the poverty and obstinacy of his materials will permit, and the winds, storms, and vicissitudes of circumstances allow—this is certainly no argument against His freedom, but is evidence of that to which perfect freedom leads: right use of freedom. The infinite multiplicity of device to effect ends in view, *e. g.*, the inexhaustible fertility and ingenuity of millionfold methods of defence in plant or animal, shows the resourceful spontaneity, the glorious extent or range of freedom. Every cell in its composition, function, and location is indeed an act of free choosing on His part, to correspond with the ever varying object in view and the modifying circumstance. Concerning God's freedom, therefore, its limitlessness is simply astounding.

Organofaction, as we have seen, is simply the automatonization and extension of function. Individualization, or individuation, is simply the unitization of organs. Many organs make an organism, or an individual. Function, we must never forget, always precedes structure; it is the repetition of function that begets that cell-habit and differentiation that form the basis of organogenesis and organofaction. Freedom is seen in the structureless pseudopod and in the developing ovum, and its characteristic is that of a seeking and making means and tools wherewith to realize its purposes. No living being can ever escape the service of the necessity of nutrition; it is always the controlling and chief element of all biologic process. In the lowest types, together with the regenerative necessity, it is the all-absorbing concern. So difficult has been the task of incarnation that, up to the bringing out of the human, scarcely any conscious function of the individual is operative except these two. And as the reproductive function in all below the human has been kept below the threshold of consciousness, *i. e.*, has been solely controlled by Him without delegation to the creature, it follows that all, or almost all, function below the human, whether conscious or unconscious, has been nutritional. Now all consciousness or individual choosing on the part of the creature is merely the delegation of freedom by the

Creator. With individuation came the individual's control of molar motion, the first control of masses or of gravity, thus indirectly gained by Biologos. But the organismal origin and independent use of molar motion, as it were, *per se*, created the responsibility and limited freedom of the individual. The molecular function of nutrition has never been given over to the creature; it is not a matter of consciousness or of delegated control; with the food past the mouth his function ceases, and His begins. But within the necessary limits, the procuring of the crude materials of nutrition (in which of course are included those of protection, warmth, and struggle) is the product of the free use of muscles or the control of molar motion. Within the limits dictated by the nutritional difficulty and necessity, therefore, have been confined the range and power of freedom, or the consciousness of the vegetable and animal. Even in these kingdoms there are millions of degrees of freedom, according to the infinite diversities and difficulties of the nutritional problem. The freedom of the barnacle and that of the fox are vastly different in degree, but that of the barnacle however restricted is never absent. In the higher animals it is very extended and rich, and in their play, humor, ornamentation, bird-song, and the like we see ever-increasing freedom, and reachings-out after freedom beyond the hard confines of nutritional necessities. In dogs and domesticated animals, the willingness of Biologos to emancipate, and the capacity to be emancipated, is abundantly evident in the beautiful eagerness and expertness with which they respond to the kindly teachings of spirit, imbibing freedom with hearty relish and thankfulness, and using it with a loyalty that often puts us to shame.

In the vegetable world to the careful and sympathetic

observer two exquisite facts are clear: the dumb hunger for freedom, and the patient self-sacrificing resignation and fulfilling of the allotted task. Every plant in its lowly beauty or conquered difficulty; every tree in its firm-rooted stability, or columnar dignity, and light-loving aspiration;—all are humbly patient and silently successful, all infinitely ingenious in seeking help of wind and bee, or in guarding against inimical self-seekers, all silently and incessantly asking of man a helpful bit of sympathy, protection, nourishment, and always quickly responsive with instant gratitude and service to him when he answers their call. Do not each and all ever speak to the kind human heart a pathetic lesson and a sweet reminder of our common essential brotherhood? For our brothers they are; God's own life is in them, and the beauty of love is that it does not even stoop to the humblest, but in its own humility walks and works among the humble, one of them, grieving in their grief, glad in their gladness, helpful when they call, and grateful when they give. These are the least of the children of the great Father, and, like all children, they seek and repay all the interest that we, their larger brothers and sisters, can give. In their humble devotion to the duty of their dual task, the perfection with which it is done commands our admiration; but, as always and everywhere, how wonderfully God rewards them for their exquisite obedience, by beauty and satisfaction. If the flower were *only* to catch the bee's eye; if the fruit were *only* to catch the bird's eye; if the chlorophyll were alone for utility, dabs and blotches of unsymmetrical and coarse colors would do quite as well as the heavenly tints and groupings. But the dear, bountiful, laughing, beautiful God is ever watching for opportunity to reward, ever trying to burst through the necessary and the useful with His rich self-revelation of the Spontaneous and the Beautiful.

The progressive delegation of freedom to the animal, as we have seen, is limited almost entirely to that degree of freedom that is conterminous with the nutritional problem. With the breaking of the animal into the human the old necessity persists but the limits are widened. Civilization is the settlement of the nutritional problem, and the permission of the application of consciousness to other than purely nutritional questions. Control of power has advanced and its other aspect, freedom, has synchronously widened. The human being of to-day, who by choice or necessity is occupied solely with questions of nutrition, with money, or the lack of it, with material luxuries or necessities, with "society" ideals and pleasures, is simply in the animal stage of existence, his freedom is that of the powerful animal limited in its range to one problem.

He has not comprehended the object of civilization, he ignores the greatest blessing ever offered man, the gift of a new freedom, of an enlarged consciousness, and of a mightier power than his worthless wealth has dreamed of. For God has ever waited to give until the responsibility could be used without too great danger to the general process. Nothing seemingly pleases Him better than to endow with freedom so soon as it may be safely given. It is no pleasure to a rider to tame a stupid, headstrong horse by the brute force of constantly-used bit and whip. The pleasure is in the bright, responsive intelligence, quick to see and loyal to act. To the loyal intelligent human heart who never abuses freedom, there comes a wealth of freedom and of power that makes the heart swell with gratitude. Whereas the selfish, the cunning, the unscrupulous, and the pig-headed find a thousand obstacles in their path, whilst enslaving and opposing enemies rise up to obstruct and to dispute advance. "Ask and ye shall receive, knock and it shall be opened unto you," is a beautiful fact of freedom and responsibility.

There is no such enemy of freedom, nothing so narrows and enslaves, as that outrageous form of selfishness, the love of luxury. Those who misuse wealth, those who ravenously seek it, and those who hate their poverty, are alike lovers of their slavery, ignorers of the freedom that would come to them if they would fill their hearts with the old simple virtues of cheerfulness, kindness, and gratitude, and look out to understand and to love instead of nursing their own deserved misery in ignorance and hate. "Magnus inter opes inops." It is the fatuity of low characters to enormously over-value what the world calls success, and a further evidence of their wretched folly consists in seeking it by chicane and fraud. For a while they rise quickly, and in the dazzle of their progress the knowing ask themselves, Is this really that sort of a world? Three mistakes are made: 1. Dependence upon "sharp" ways and means lessens one's knowledge and recognition of honorable means; the real sources of power, intellectual clearness and honor, are ignored, whilst the love and respect of high-minded men slips away into silent, subtle distrust. 2. With temporary success, one grows more brazen and the natural limits of trickery are not seen. Soon there is collision with an immovable obstacle in the shape of honor and morality incarnate in some man or institution. 3. If united to a sufficiently strong intellect "shrewdness" and selfish cunning may be quite early successful, but whether so or not, the attainment of the long-sought prize turns out to be a sad deception, the ashes of death. Many a successful man or woman would give all their riches and luxuries and power for one fresh heart-thrill of large joy, for the *possibility* of being kind, or good, or intelligent. But dead forever is the murdered heart, the stifled honor, and the starved intellect.

The degree of freedom is therefore according to the desire and ability to make right use of it. Every mechanism and creature of the living world,—that world-process itself from beginning to now, is evidence and fact of deputed power. Nothing is plainer than the present desire to still further delegate responsibility so fast as trustworthy delegates may be found. Here stands the splendid completed engine of the entire biological world, together with command already gained of large mechanical powers of the inorganic world, a gift to man by its Creator. That huge fortunes are being made by individuals with its use and abuse; that city luxury and vice are expensive parasites upon country hardship and virtue; that governmental rule and legislation are used as the tools for the greed of classes and even of individuals; that forests are being destroyed, the balances of different orders of animal and vegetable life thereby upset, and injurious climatic and agricultural changes thereby caused; that some of the fairest of God's creatures are being barbarously and ruthlessly exterminated —all these things and many more show that responsibility and freedom have been delegated to man too bounteously and too soon, rather than too niggardly and too late.

I said delegated, but here arises the thought that it has been a ruthless seizing rather than a receiving of gift. Freedom is doubtless an acquirement, a method as well as a reward of education. By the wrong use of it one lessens the right use, and already we see the subtle mechanisms of forefending, compensation, education, and equilibrium being instituted and set to work. These mechanisms are everywhere coming into play, and are of most various and multiform nature. Preventive medicine; societies for the prevention of cruelty to animals and children; the science and art of forestry; the sympathetic study of the living world, great Science herself; astronomy; democracy; the growing detestation of selfishly-used wealth; and socialism,—such are a few of the agencies at work. Few know whither they are looking or for what they are working. God makes agents and instruments even of His enemies, and, our compasses having been lost, with His own winds he steers our vessels into harbors unexpected and better than those we have sought. But by the voyaging we have learned the sea, and a new life, and we are henceforth better mariners than we were before.

Finally, there arises the question as to possibility of any thing that may legitimately be called freedom, consistent with the biologic process of "incarnation." Every cell of every organism being made and fed by His life, how is it possible to find in the very work of His hand anything but a mechanically obedient mechanism, and where is there chance for freedom or responsibility? The answer to this very pertinent question is clear to my mind, but it is difficult to put it in words, and almost impossible to do so for those who by mental habit have anthropomorphized their conception of God in such a way that the truth becomes impossible for them to see or vividly to realize. The question is bound up with that of the nature of spirit and personality. There is also a certain intellectual expertness and power requisite to catch even the faintest perception of the real nature of spiritual existence and life whether of the divine or of the human type. Occidental thought has had no gymnastic exercise in comprehending metaphysical realities, and it is either awed and bowed into the dust before God, or stupidly denies His very existence. But whether adoring or denying, it is non-understanding. Then, too, a very general idea and ideal of freedom is that of freedom from responsibility, instead of which, freedom *in* responsibility is a much truer conception. The only true freedom consists in obedience, not in disobedience, to use power rightly instead of wrongly. The analogy of the horse may serve us again; an unruly, balky, stubborn animal will secure from his owner less freedom than a bridlewise and gladly willing one. One that would never be caught or be broken would finally be killed. The more "vicious" the more certain to get suffering, and to drift to the worst and severest labor. The more humanized, the more freedom and love; and, if intelligence and loyalty were sufficient, restraint, bridle, whip, and all bars to freedom would be done away with. I find it exactly so between God and His creatures. Joyful and intelligent acceptance of the rôle and duty assigned, courage to be and to do, lead to a freedom no money-king or man-king can ever have or can even understand. Loyalty and knowledge combined are the only means of acquiring spiritual freedom, and these things are impossible while money or power is selfishly used. The very acquiring and holding of wealth is itself unforgivable sin and wrong,—is evidence *per se* that self has been valued above all others and above the purposes of God. The words and the truth of the words of Jesus are perfectly plain as to the love and use of money, but no modern Christian (unless perhaps it be the "heretic," John Ruskin, one of the greatest characters of these centuries) is at all likely to heed or to follow the command of his God, the "Second person of the Trinity." A worshipper who in the week spends six days in the porcine scramble for stolen money, and seven evenings in wasting it in self-gratification, who lives or bends all energies in trying to live in affluence, and who gives one hour out of one hundred and sixty-eight to the worship of a deity whose thundering repetitive denunciations of money and the love of money are the plainest of His teachings,—this is a spectacle reserved for the blasphemous degradation of the nineteenth of Christian centuries!

CHAPTER XI.

PERSONALITY.

THE telephone, the telegraph, travel, and increased ease of travel, knowledge of geography, astronomy, spectrum analysis, the study of gravitation, radiant heat, light, electric and magnetic ether-motions,—these, in one sense, are athletic mental exercises in space-comprehension and in space-subjection. We are no longer appalled and terrified; instead, we understand and use. Mere distance, however interesting, no longer inspires us with cosmic horror. There is no comprehension of the world of lifeless matter or of the work God is doing in and with it, except through a diligent and an accurate study of physics, molecular and molar. These are the lowest stairs that lead up to the temple of the knowledge of God. As commonly understood worship is proof in itself that the Being worshipped is not known; it is often a lazy excuse for not trying to know Him; and even in its highest aspects, it is only a combined admonition and humble non-actualized desire to know. Mere blank wonder, mere rapt adoration, mere dazzled awe, mere non-comprehending averment of God's existence and greatness, mere emphasis of the creature's littleness, mere flattery and praise of Him, however solemn and sincere, is, to be sure, infinitely preferable and more gloriously true than the prize stupidity and despicable *nil admirari* of fashionable agnosticism and atheism. But, frankly, it also is at last a little *bête*. Certainly there are degrees of knowledge. Even the priests of the God *Unknowable* would hardly deny that they themselves know a wee bit more about Him than an African pigmy knows about Him. Every church and sect is quite certain that its peculiar worship is more right and is based on a better conception of God than that of the other churches and sects. All do tacitly admit degrees of knowledge. If God exists, what thing could He more desire than to be understood and known by His children. Instead of impiety, the attempt to comprehend Him is true piety, and to increase and praise the open-mouthed bewilderment of stupid contemplation is genuine impiety. Even a good man dislikes the over-enthusiastic praise of his admirers, however sincere it is. Certainly a good God must smile pathetically and tiredly upon the " hocus-pocus," and He must long for the coming of virile human minds that, without losing love and respect, will still try to comprehend and to know.

The knowledge of chemistry, of molecular and atomic physics, and especially of microscopy, also complements and completes the knowledge of the infinitely great and far, and habituates the mind to a just comprehension of the idea of extension. Knowledge of the ether and of its properties has annihilated nihilism, the old bugaboo of emptiness or vacuity, so that now we know that wherever space extends it is filled with a medium that binds the physical universe into a unity and comprehensibility greatly aiding our sense of home-being and home-making in the world. Waves of water that we can see and hear, help us to understand waves of air that we can hear but cannot see, and waves of air help us to comprehend the attributes and waves of ether that we can neither see nor hear, and that are studied only by the scientific imagination. Rising from that completed study the mind is trained to grasp the idea of an unlocalized intelligent force, still less subject to the rule of space or of extension, and operative throughout the universe. The uneducated mind demands localization of spirit, and feels it is deprived of its divinity if it is forced to think of Him as unlocalized, or as at all points of space at all times. When that intelligent force enters matter it becomes subject to its laws, and as regards both action and thought there is imposed upon that specific organism the primal condition of the law that no body can occupy two positions at the same instant. Hence the necessity of our thought (at least temporarily) is to make us localize personality; and if it be not so localized it becomes almost or quite non-existent. Just here the simple perception of a fact corrects our loss of comprehension, and gives us back the reality of God by the reality of apprehension. At any one instant and at all instants we perceive and know He is actually forming and upholding every cell of every living organism in ocean, in air, and on earth, all over and all through the world. Our own attention must be upon a single object at any instant; His is equally strong and accurate upon billions of millions at the same instant; and not only in this little world, but in others throughout the depths of the universe. The fact we can perceive; it is beyond question; and there is also an ability on our part to grow into a progressive comprehension of the fact, as we habitualize and progressively strengthen the mind to such perceptional function.

In the same way we can accustom and exercise the mind into a conception more or less true of the nature of time. A knowledge of history, human and biological; the conception of gravity, and of swift-flitting light; the vivid memory of our dreams; the realization of the unity of end, middle, and beginning, of our life-time; the comparison of sluggish with brilliant minds; the evident compression of time by the life of a modern educated mind,—such things habituate us to the thought that time is an attribute or condition of matter put upon the incarnating spirit. It is in the last analysis a sort of measuring-stick of the nutritional difficulty. The biological process is so slow because of its difficulty; if nutrition were a hundred times easier, our thought and our life would be infinitely more rapid, and the present stage of progress would have been reached long ago. Education and scientific thinking in a measure annihilate time, and the timelessness of God becomes to the expert intellect not a frightful but a comforting conception. Fact, the simple perception of an evident fact, again aids us: One purpose and manner of work unites the earliest efforts of the incarnation-process with the middle that we are, and with the coming perfection of that that is to be. There is here a unity that could only flow from a cause lying outside of time. The length of the unrolling is the condition of difficulty of the process, and the length hides from the undiscerning eye the design, patent to him who can overlook and extinguish time in the perception of cause and effect. Trained thought enables us to squeeze time out of fact, to compress centuries and eras into minutes, and to see that to God the process is instantaneous. To the slow-travelling mind the outlines of the design of the biological process are blurred by the dreary journey, but he whose spirit can travel by the wings of the educated imagination, crushes cycles of time-changes into seconds of thought, and walks boldly out of the opened prison-windows of time into the freedom of an eternity wherein past and future are unified into the ever-present.

The limitation of God's omnipotence comes chiefly to light in the nutritional difficulty of the cell and organism, shown in death and in the impossibility of reaching ends (the human co-worker, for example), except through the millions of precedent and preparative stages that we see in the earth's biological history. The human aspect or generalization of that difficulty is our mental habit of seeing things only in time-relations. And as that difficulty of God was tremendous and incessant, the record of it engraved in our minds is powerful and indelible. God's space-difficulty has been slight or non-existent, and there is a similar freedom of our thought from this slavery. Light takes hundreds or thousands of years to travel to us from some visible stars, even though it travel at the rate of 189,213 miles a second. But our thought can travel from *here* to *there* instantaneously. In the same way our practical reduction of space by quick travel, by telegraph and telephone, has been tremendously successful, whilst life is not lengthened, and our control of time is but little if any increased. In many such ways we see that God's great difficulties are also ours, and that His nature and capacities are also our own, with the obvious difference of degree and perfection.

God's power to direct attention to all parts or points at once is a natural consequence of his practical omnipresence. Our development and history have been such that it is quite impossible for the inexpert mind to fix clear attention upon more than one object at a time. The organist or locomotive-engineer has quite a high power of divided attention. The ability to hold the attention so aloof from any one point, and yet so alert and easily influenced that disorder in any one part will quickly direct the attention thither, is observable in many, and this power is doubtless the beginning of the process of our infinitization, if one may so speak. Immense progress in carrying many subjects "in hand," is by slow degrees leading the complex modern mind·into the capacity of multiform synchronous attention. It is a distinctly improvable power, a matter of mental gymnastics and exercise. If the Brahminic and Buddhistic mind, with its splendid metaphysical suppleness and power, had but had the actual world of matter and life, the scientific data of the modern, to work upon and to balance it, its mooniness and pessimism would have been obviated. There might also have been evasion of the distressing narrowness and banality of the modern type of pseudo-science, which, as it were, is forever content to study mysteries profusely and strangely flung at us from over a wall, but that never dreams of asking about the other side of the wall.

The next most striking quality of the divine mind, one that is perhaps more noteworthy and characteristic than any of the preceding, is that of His ability to deal with atoms, molecules, short ether-waves, and the infinitesimally small. This is conjoined with that striking limitation of his omnipotence that I have so often emphasized, His inability to directly influence molar motion. The entire fact constitutes the most astonishing and marked difference between the divine and the human personality, and, if we would understand the living world, needs to be kept constantly in the attention. If one wishes to think of God as "substance" or "matter," it may be done, and the knowledge of the ether greatly aids us in the attempt. According to this suggestion we are helped to imagine

His omnipresence by the thought of an infinitely fine fluid or body, like the ether, infilling all space. His freedom from time-limitations has also a crude analogy in the inconceivable rapidity of ethereal motion, by the continuous wave or mass never still, and also by ethereal permanency and unity.

It is an interesting fact that, of all material substances, the smallest, thinnest, slightest, and most physically unknown of all—the ethereal wave, is the only standard of measure we have that is fixed and always and under all circumstances the same. All other standards of weight and measure are ever changing, and ever liable to change. Is this a hint to us that the cruder the "substance" the more untrustworthy and the less "standard" it is, and that the substance we call God or Spirit is at last the only absolutely durable and standard substance in the universe? The more the density, the more unlike God. The hint is proved by noting that substances of all kinds are valuable to us in proportion to their smallness, fineness, thinness, or gaseousness. The most useless thing is a dead planet of crude inorganic or mineral matter. The dead masses of rock that form the bulk of our world are quite as useless and non-significant. The pulverized rock or "ground" forms a rooting-place for plants, and helps to furnish them an equable supply of water and a few constituents of nutrition. Water is another release from hated solidity. Although it forms some four fifths of our bodies, its rôle is not vital; it only forms a medium of communication and of assimilation. It is a good slave to bring and carry the letters of life, but it may never read or indite them. It is the atmosphere, however, gaseous and so near "nothing" that the savage does not discover it, that feeds plant and animal, and is an instant, absolute, and vital necessity of every leaf and lung. With this attainment of "thinness," also, begins the psychic life and intermediation—the necessity and spirituality of sensation. The atmosphere furnishes the material basis of sound and music, and all that follows therefrom. The mechanism of the biologic process is nutritional, and all nutrition consists in the use of atomic and molecular forces. Of these man's mind has no direct or sensational knowledge; only by inference and mathematical reasoning do we know them to exist, or know any thing about them. It is certain that no eye, by whatever microscope aided, will ever see a single atom, even of the largest variety.* These atoms and molecular motions, however, reach consciousness through the senses of touch, taste, and smell. In the ether, however, we have the most valuable and essential of all physical agents. It is the medium of temperature-regulation and supply, by which and in which every nutritional and sensational function subsists, and persists, whether in plant or animal. Heat is the condition, index, and register of all forces, of biologic forces especially. As the ether is millions of times more thin than air, so is it millions of times more vital and essential, closer, and more nearly in contact with our being. As a light wave is millions of times smaller than an air wave, so is vision proportionally richer and more psychic than hearing. The letters of the alphabet are conventionalized pictures, and on the physical side all human intelligence is the product of vision.

* The smallest living particle visible under the microscope contains about two million somacules (living molecules). About 85 per cent. of every protoplasmic mass is water, and a large part of each cell is not true living matter, but is either food or dead waste product or excreta. Hence the real directing and modifying part of the cell is composed only of from ten to twenty million of somacules.

The inference draws itself: the "substance" we call spirit, life, or God—Biologos,—this subtle fluid, as infinitely removed from the crude ether as that is from air,—this is the life and heart and essence of us,—it is the ego itself.

I wish here to draw an inference of analogy, and to fix it firmly in the attention: I believe the fact that God can only affect matter through the infinitesimally small forces that we call atomic or molecular, is from precisely the same reasons that radiant heat can only so affect it, and that His causative agency is accurately analogized by the action of the ether on vibrating atoms. It is an easily understood mechanical law that in order to move a body a force must be proportional to the resistance or "weight" offered, and if the body to be moved is vibrating or oscillating, the instant of impingement must be timed to the phase of the vibration. It is in accordance with this necessity that ether waves can only affect a mass by affecting the constituent molecules or atoms of that mass, since they alone are small enough for so small a blow to have effect upon; and furthermore, their vibratory period must be adjusted, synchronous, or nearly so, with that of the vibrating atoms. One further observation is yet necessary to make clear this conception of what may possibly be the mechanism of God's influence upon and use of matter: The ether is thrown into wave motions of all degrees of rapidity or wave size. Waves much smaller than those of the ultra-violet end of the spectrum have not been much studied, though there is hardly any limit to their creation or existence. But down through the light spectrum, and for a dozen octaves below, the longer waves have been measured, and lately Hertz and Tesla have shown us that this limpid medium may be thrown into vibrations of any length desired, these longer vibrations constituting the phenomena of electricity and magnetism. So that, with the same medium, there may be produced waves a thousand miles long, and from this point all the way up to those of perhaps one one-hundred-thousandth of an inch in length. I argue from these facts, and especially from the fact that Biologos can only control mechanical forces so small as those of atomic and molecular vibrations, and that these He *does* control, that His "substance" or "body," inconceivably more fluid than the ether, influences atomic and molecular vibration by its synchronous and dominating vibrations. It is only the largest and strongest of His vibrations that are sufficiently powerful or adequate to the task of influencing the fine small motions of our known atoms and molecules at the degree of temperature of organic bodies. Behind and beyond materiality proper, or materiality as we know it, the extension of activity of the divine Being within Himself, by finer vibrations than we can know, or by other undiscovered modes and forces, leaves room for a superhuman imagination to play in, and for the inexhaustible riches of His nature.

I see no reason for shrinking from such conclusions or from such methods of reaching conclusions. It is more truly reverent than the pure blank, non-comprehending wonderment of the bewildered worshipper, and it is more courageous and loyal than the lazy cowardice of agnosticism and atheism. The danger to any but strong, clear thinkers is that any endeavor to de-anthropomorphize their deity seems to dissipate Him into "thin air." We are so material, and so crudely and clumsily material, that we have little trustfulness in this "thin air," of the spirit, but which alone has stability and permanence and use and beauty, whilst instability, impermanence, uselessness, and beautylessness increase with every increase of density. To the soul the cry of religion and of science,—ay, even of physics, is, Trust your wings, the air is your home!

God's perfect kindness is shown by the existence of the wings, and also of the muscles to use them, when the idea and the desire of flying has come. Those who have no spirit-wings will find such ideas ridiculous, and dangerous desire will not be aroused. Such must longer rest quietly in the nest of practical materialism or materiality, fed yet longer by the kind mother-love of the senses.

But to those who do trust this upbearing air, there comes at once the consoling thought and the repeated perception, that, though not localized, though freed from the binding of time, though always here and there, though attentive at all points and at all times, though most unsolid, He is yet Person. No true element of personality is lost, either to the pure reason or to the physical eye. And most contenting of all is our likeness to him,—nay, that in very fact and truth we are His children, transplanted and emigrated parts of Himself, set to the task of breaking and tilling a new soil which has been provided for us, and for which He has outfitted us. Intention, foresight, thoughtfulness, and a thousand thousand signs of Himself as kindly, helpful, intelligent, and purposeful, are scattered wherever we come or go. To the spiritually erudite eye, the organic world is a palimpsest, and beneath the beautiful modern writing of biology, there gleams in dimmer characters the significant and beautiful revelation of an ancient wisdom and the story of divine passion.

Perhaps there is no characteristic of God that so clearly proves personality, or that more closely allies us to Him, and that more boldly and repeatedly commands our sympathy and delight, than that of His ingenuity. If we realize in thought a fraction of His difficulties, and if then we seek to penetrate the secrets of the intricate mechanism of any one organism or fraction of an organism, we are astounded at the inexhaustible fertility of resource, the infinite subtlety of cunning, and the perfection of ever-varying device. All of this, too, with an undiscoverable deviousness and indirection forced upon Him by untoward material and circumstance. This recalls the charm and pleasure of mechanic invention, discovery, and use in our own world and work, and shows us whence we derive our power, whence come our use and pleasure in the power. The kinship is clear. A great mistake of the religionists has been to think of God as a far-away, a huge, and solemnly mighty being, hurling worlds and great forces. But the most salient attribute of all is that of His exquisite artisanship and of His mechanical genius. Before all things is He a worker, and an infinitely deft and expert one ; with a patience of constructive talent, a perfection of detail, and a consummateness of result, that are the model and despair of every human workman. God has nothing to do with, and has no power over, worlds and giant forces. His is not the thunder, the storm, the planetary movements, nor any molar motions. He deals only with the infinitely small, but his mechanisms are the marvels of perfection that we see.*

* As one of numberless instances, take this extract from a hunter's journal : " Who could not tell a loon a half mile or more away, though he had never seen one before ? The river was like glass, and every movement of the bird as it sported about broke the surface into ripples that revealed it far and wide. Presently a boat shot out from shore, and went ripping up the surface toward the loon. The creature at once seemed to divine the intention of the boatman, and sidled off obliquely, keeping a sharp look-out as if to make sure it was pursued. A steamer came down and passed between them

And over all, in all, and through all, the smile and the flush and the flashing of a divine beauty that thrills and charms,—but that eludes!

And from what other source could our personality come except from Him? Can the lesser create the greater? A literal incarnation implies, however, that we are not creations, but that we are creators. All our work in this world is in small exactly the same work that God's work is in large, and all our powers and capacities are also His powers and capacities, seeking the extension, the perfection, and the scope of His. All physical progress is approximation toward His manipulative skill, and all spiritual progress is approximation toward his mental and moral and esthetical perfection. Any true progress whatsoever is progress in Godlikeness. But only a large and deep study of biology and anthropology can define what God is, and in what likeness to Him consists. We are God deputed to the work of incarnation, and in entering matter God must devote attention to the work in hand; hence the temporary restriction and limitation of our minds, their finiteness and groping, ever clearing to a more definite consciousness of essential unity and identity with all spirit. Our

and when the way was again clear the loon was still swimming on the surface. Presently it disappeared under the water, and the boatman pulled sharp and hard. In a few moments the bird reappeared some rods farther on, as if to make an observation. Seeing it was being pursued and no mistake, it dived quickly, and, when it came up again, had gone many times as far as the boat in the same space of time. Then it dived again and distanced its pursuer so easily that he gave over the chase and rested upon his oars. But the bird made a final plunge, and when it emerged upon the surface again it was over a mile away. Its course must have been, and doubtless was, an actual flight under water, and half as fast as the crow flies in the air." Think, also, of the physiological and neurological mechanism of the flying monkey that can rush through high tree-tops faster than a man could run on the ground, yet " with his head turned back and his eyes fixed on his pursuer."

minds are temporarily weak and warped from the struggle with matter; we are more or less forced to see things in time and space relations; the all-ruling necessity of nutrition is on us; we are compelled to singleness of object in attention; to dealing with solidities and masses and molar motions; but every day shows us progress in freeing our minds from these bonds, and in infinitizing our mentality. In very truth, the genuinely personal is hindered by these bonds and inherited experiences, and true personality is that of God, not that of the spirit subservient to materiality. We are coming to personality more and more, but it is only as we cast off that which the deeply materialized mind mistakenly holds as most personal.

I here catch glimpses of a divine tragedy that was symbolized and exceptionally seized in the Christian theory of the incarnation of Christ. The conception of the Infinite Father leaving the throne of Heaven, and of Christ taking upon Himself the human form, burying Himself in matter or flesh for man's sake,—this is all a prefiguring and exceptional *Ahnung* of the world-process. And as Jesus gave the plainest evidences of his possession of the limitations of our common humanity in the real suffering, mistakes, and misconceptions of his life and teachings, so in the great incarnation of Himself in the biological world, God resignedly blinds Himself to His more infinite characteristics, and in devoting Himself to the huge labor, He temporarily renounces their functions, and sees with the human eye all things as we see them. This does not necessitate a limitation of His infinity, the permanent or absolute inability to devote attention without becoming "absorbed" in his special work, but would imply simply a postponement of useless function until its play or activity shall not detract from or endanger the hazardous process in hand.

In devoting Himself to the work of incarnation, God, as it were, rendered himself temporarily unconscious, forgetting for the time His higher attributes. The higher consciousness and the fulness of the divine characteristics would have been hindering and disturbing elements, in the early stages at least, of the limited function. In plants and animals He is thus perfectly "absorbed" in His work; but in man, even the savage, the hint and suggestion of the divine is ever seeking clearer ways and workings, whilst in civilized man, Godhood gleams out upon us from all history and from all great personalities. In the future the God in man is coming to perfection; or, in other words, God resumes His fuller consciousness and powers, while at the same time utilizing His new acquirement,—control of molar motion. Postponement and temporary renunciation, made necessary by the rigor of the nutritional difficulty, appear on the human side as the temporal, spatial, and other limitations and imperfections of our mentality. The end of that postponement is the solution of the nutritional difficulty through civilization; the sudden blooming of science and art is the beginning of the fact of once again taking up His disused powers, and of accustoming Himself to entering upon His old-time pleasures and exercises. The infinite Son is returning to the home of the Father with the consciousness of work well done and reward awaiting. The suddenness of the present blooming of science and of strengthened mentality shows how natural is divinity to us, how promptly we can catch up our old work, even after a million years of disuse, and of devotion to other tasks.

But, with unconscious truth the old creed said, "not two Gods but one God." Father is Son; Son is Father. There is no genuine divisibility of the divine nature.

And, moreover, the return to the Father is not renunciation and leaving of the former work. The true Buddha never enters heaven until the last of his brothers has entered before him. In other words, heaven is not *there*, but is here and now, and no work is ever quitted. The process is still going on, the Son is ever returning to the Father, because the Father is here, and progress so far as we can see is infinite.

And finally, this plainly leads to the foreseen and glorious conclusion, that the Son is not single, but that we are all His Sons. Sin, suffering, slowness of progress, disease, death, indirection,—all these are the expenses and difficulties of the process; whilst stupidity, with all forms of mental obscuration, are the temporary effects of temporarily absorbed attention and renunciation of higher life. If incarnation, as is certain, is a literal truth, then we are not only sons, but the God Himself. Every possible thought or system of intermediation, of a hierarchy of gods breaking the fall from the infinite to the finite, or constituting the steps of the stairway from the finite to the infinite, all must end in deserved and miserable failure. Because we have not comprehended the finiteness and knowableness of God, or because we do not recognize the infiniteness and unknowableness of man, we baulk at the thought of the literal identity of the two. We think of ourself as a body, but to the last atom our body is our instrument. We are the supercorporeal being that directs the body, uses it, receives by it, moves it and effects purpose with it. And this true ego is absolutely identical with the Ego that infills and stands behind all living forms, lending them life, and making them serve purpose and mentality.

Here, then, at last, is the final solution of the problem

of freedom and necessity. By no other theory that I can conceive is it possible to slip the bondage of necessity and determinism. Unless we be really possessed of the divine freedom by the fact that we are sharers of the divine nature and being, we are only self-deluded slaves. Any sort of subordination at once throws responsibility back upon the chief or highest authority, and thereby automatonizes man. So long as man's foolish thought made him desire an impossible sort of freedom, that without accountability or responsibility, just so long was true freedom impossible. We have wished to be irresponsible tyrants instead of constitutional governors. God Himself has no freedom of that sort. Freedom is not only, and is not so much, release from restraints as it is choice of right and necessary restraints. We are possessed of exactly the same kind, and of a progressive degree, of the same freedom as has God. His freedom, so far as the world-process that we know is concerned, consists in utilizing every means within His grasp and power, to incarnate Himself in living forms to the greatest extent possible, to humanize and to give to the human the control and use of physical and subordinate biological forces. Our freedom consists in the recognition of our true origin and nature, and in accepting and utilizing the responsibility and the power as co-workers with the divine, and as sharers both of His being and of His task. We are God involved in the work of utilizing waste mechanical forces, of spiritualizing matter, and of vivifying dead worlds ; and with it all, educating and perfecting spirit, learning the ways, the necessities, and the uses, both of matter and of spirit, and of that new world arising out of the jointure of the two, the incarnation-process. Your non-recognition, your rejection, or your misuse, your faultful or selfish doing of the duty and work, is both your failure and His failure, is His mistake and your mistake, your loss and His loss; you cannot escape each other, for you are He and He is you. With loyal loving adoption of His aims and work as those of our own, we advance to a realization of the highest personality and freedom, awe gives way to love, and difference passes into identity. The shadows of things, and the obscurations of materiality are lifted, the wings of the spirit are grown strong: *à Dieu !*

CHAPTER XII.

IMMORTALITY.

PEOPLE generally cherish the delusion that their hunger, pretended or real, for immortality is a virtue and a thing to be proud of. Careful observation has convinced me that in many cases it is distinctly and nothing less than a very narrow sort of selfishness. It does not usually flow from a large love of life, or of the things Life is seeking to bring about in this world. People who believe most vindictively in the belief are such as have done very little toward enlarging their own life or that of their fellow-men. It is commonly supposed that there can be no greater heresy and injury to both religion and morality, than any negative attitude toward the belief, or doubt thrown upon it. But I am convinced that it is very commonly, if not usually, of distinctly pernicious influence upon character and society. If one believes in the old-fashioned "soul," and in its "salvation," and that one's own soul has a surety of heaven, nothing can more effectually breed practical fatalism, *laissez-faire*, and egregious conceit. The consummate ludicrousness of a "sanctified" and saintly crank is only better concealed in many who have been less clumsy in the self-cheating and delusional processes of mind whereby they have erected a very high wall to hide ugly truths and plain duties from view. The perfervidness of the belief, moreover, has always a plain smell of intoxication about it; it is almost always an artificial emotion, whose strength is largely dependent upon the amount of misery and poverty hidden by the illusion of a false gaiety and a pretended certainty.

There was once a splendid scoundrel who defied government and armies with a horde of invincible slaves to whom danger was delight, and who sought death by unfaltering obedience to their master. Their belief in afterlife, and in its disposal by their tyrant god, was made incomprehensibly strong by the trick of drugging them with a narcotic, and, while insensible, conveying them to a mountain paradise where every delight of every sense was drenched with satisfaction. After this foretaste of heaven, they were again put to sleep and conveyed to the world, and after this there was no doubt about obedience to the commands of a master who had at his disposal such a heaven as that. To get on the safe side, to win by excess of flattery or obedience the good will of a tyrant god, has been too common a characteristic of religion; and so, alas, has the seeking of immortality often been a sharp looking out for "number one" in the chances of life. Just as the sacrificial aspects of much religion, scape-goat sheep and scape-goat Christs, have been tricks, fine or flimsy, to get rid of conscience and compound with the devil called God, so the attainment of heaven has frequently been a fine game to get the advantage of one's enemies, and of those not so cunning. The rabidness of the belief has usually depended upon the proportion of the few saved and the many damned, and the frightful immorality of any salvation whatsoever enjoyed, while there was any damnation suffered, was a thought kept well out of sight.

The unconscious power and origin of the belief, however, have, of course, come from larger and deeper minds and reasons than such self-seekers could fathom. The arising of the belief in historical times and religions, where it was heretofore non-existent, seems to me another example of the breaking forth of the consciousness of man's divine origin in nature, this consciousness and belief having been kept in abeyance by the demands of the nutritional struggle and progress in incarnation. These early shapings of the doctrine are the crude ploughing and breaking of the ground for a better harvest. No great religious truth comes so suddenly to perfection as this, and the belief must yet cleanse itself of outrageous crudities, and perfect itself to finer issues and more truthful truth. After the establishment of the fact comes the limitations of degrees, and the refinement of qualities. Like all other deductive and religious truths, this belief must be made scientific; the rational systematizing consciousness must take it in hand, and it must submit to the calm estimation of proof, limit, and degree, of a sympathetic but unbiased judgment.

"If a man die shall he live again?" The very wording of the question betrays the visible hope, the doubtful longing. One feels the wavelet poised for an instant in semi-independence, and the large ocean certain to draw it back again. The wording also betrays the crudity of the conception, and answers itself. There is shown an entire lack of discrimination as to meaning of words, and extent of facts. What is man? What is death? What is it to live again? To the unthinking the answers seem very easy to give, but to the thoughtful they arouse profound counter-queries.

In the ordinary and accepted definition of the word, "a man" is understood to be any representative of the genus *Homo*, whether he be a nameless African savage, or one of civilization's most cultivated scientists. If the fact of possession of the human form implies the possession of an undying soul-life, then no line can be drawn at the upper limits of the animal or vegetable kingdoms. Everything that lives is so strikingly filled with the same mental force, that it is the veriest trifling to deny any animal and plant the same right and necessity of future living as ourselves. This brings into view the clear evidence of absurdity in the common doctrine. According to that doctrine, heaven, or the function of the future life, is to be one of enjoyment and general objectlessness. There is here a withdrawal of divine energy from use or work that is unlike anything else known or thinkable in all the worlds of mind or matter. If the possession of such "soul" divisible from body is certain in man, it is, as I have said, certain in animal or plant, and if by natural or divine ordering, this homo-soul, at the death of the body, rushes away from earth and work, so, precisely, must animal and plant-soul spurn matter and life therein. Not to assent to this introduces a new principle: either animal-soul and homo-soul are essentially different, (which nobody would now be silly enough to affirm), or there is introduced a stage of progress or a degree of soul only thus rewarded. To make salvation or the heaven-getting of soul depend upon any such indeterminate and indeterminable point of progress or merit is, from that point of view, to land the whole question in an immovable cloud of vagueness. To drain off from earth and from functional activity the souls of plant, animal, and human beings that die in one year alone, would require an infinite inexhaustibleness of the Source of Life. There is not the least reason to suppose that the law of the conservation of energy is not as rigidly applicable to Life as to any other form of force. To attempt to conceive it otherwise lands us at once into unthinkableness and nonsense.

We therefore see that any fact of immortality must be in harmony with the fact of the life or soul of all living things, and that functionless life or soul is quite as abhorrent and unthinkable as functionless force of any kind. The beauty of such a conclusion is the proof that the sort of a soul and heaven commonly desired is as inethical and irreligious as it is unscientific and impossible. "Salvation" was a spoiled child's theft of the cake of happiness and hiding in the garret of heaven to eat it alone. There may be such a heaven of inactive enjoyment and selfish pleasure, but God's Buddhas do not enter it while the struggle of the world-process is still going on, and I have such a firm conviction of His loving justice that if His Buddhas beneficently postpone it of their own will, selfish laziness will hardly be able to steal past His inattention. Parasitism is, after all, a very small fact in the world, and even that is usually made to confer indirect blessing. The radiation of the sun's light into space is not lost or losable, and if there is such an irradiation of Soul into the regions of space as the common idea holds, I suspect it is none the less functional and utilized somewhere and sometime. As neither can be annihilated, they must go on until "absorption" does take place. What is the influence upon character of persistent enjoyment or resting, is plainly seen in the depravity and mental vacuity of our social do-nothings. Life is a force, and Biologos is of all things a worker: it is hardly probable He will excuse His forces from activity, or His souls from work. Heaven is quite the last and absurdest thing to think of.

As we daily see, there are two ways whereby soul-activity is kept persistent, and immortality really attained: by heredity, and by spiritual influence. The first is more physical in mechanism, and seemingly the more powerful; but the second, I doubt not, is, in a large way, quite as real and more effectively strong. Marcus Aurelius sends his thoughts, the sample and ideal of his soul, down the centuries, and procreates spiritual children wherever his words go. But by heredity is the more certain and methodical manner, and no sin is greater than that of large minds and hearts refusing or ignoring the duty of child-raising. It is towards this that the world-process has struggled for a million years, and the thread of purpose, tirelessly and patiently followed through the long labyrinth of development, is thus cruelly snapped in an hour of weariness and waywardness. If there is any significance and object in the incarnation-process, it is bound up with the best type of civilized manhood and womanhood. Therefore, the very acme of sin against the Holy Ghost is refusal to perpetuate that type. If the old idea of a judgment-day were a truth, the first question asked of the civilized sinner would be not as to murder or any crime against present society alone, but as to the more heinous crime of disloyalty to God Himself, to His work, and to the future, by wilful disobedience to the second law of the incarnation-process.

And if by circumstance or accident child-raising is refused to one, it behooves him to devote the same energy and self-sacrifice to the fundamental aim of Biologos: orphans are to be raised, and the work helped onward as evidently purposed. Or, if perhaps to such be given an exceptional power of thought, or other means of after-death influence upon the world, that function should be exercised as a distinct atonement. "It is dangerous to be believed," only when the belief inculcated contradicts the evident purposes of the subtle wisdom moving all life. There is never much danger in simple kindness, and much of our misery and sin come from lack of it.

The genuine "victory over death" and the grave will come only in the persistence of continuity of the pliant and progressive individual. In the meantime, the thoughtful kindness of God is shown in the peace and silent ease with which duteous souls slip the bonds of finished life. To those who have completed their clear duties, death's visage is no horrible one, and to them the final rest comes as the night's sleep, which is indeed a daily warning and reminder of it, and initiator into it. From His obedient animals He has hidden the very knowledge of death, and at life's proper ending His obedient human children have become so wearied with the long day's work, and the twilight of life's evening so softly and sweetly passes into the night of death, that the silence of the unseen stars only equals the silence of the tired-out heart. The body has not been their chief love, nor things corporeal their sole concern, and, in losing the body, self has not therefore been lost. But it is they who have not been loyal to God and aidful in His work, they who have not found, explored, and utilized their own spiritual natures, these are they that fear death and shudder with horror at his approach. Reliance upon the spiritual generates that confidence in it that robs death of his terribleness. They who have loved life for the body's sake, thinking corporeality the end and not the means, such naturally feel that in giving up the body, all is given up. "The lack of evidence of immortality," of which the complaint is common, seems to me a most wise provision on God's part. The lack must continue until the task to which we are set is completed, and the consciousness of one's own immortal nature is so clear that evidence of the common sort would be positively needless, or even repulsive. It is well that the evidence furnished by the motley crew of spiritists is so valueless and inconclusive, else the race in possession of such a certainty would become as spiritually degraded and vapidly materialistic as these precious folk. God would indeed be cruel to provide such a heaven for us, or to make us so convinced of an undeserved immortality, that the lesson of corporeal life would be utterly misused. Mankind are always seeking and making for themselves the illusions of fatalisms and certainties which excuse inertia and obviate the pain and fact of progress. None such is more potent for evil than this of an unfaltering belief in immortality, which, even if the desired were obtainable, hides duty too deftly, and neglects that preparation of spirit here that alone would make a heaven anywhere. The doubt and the doubtful hope arouse the startled heart to a study of the conditions of immortality, and keep it plastic to the ideas and influence of the spirit of God ever drawing us subtly and kindly to the life of the spirit.

A little reflection will show us that what is usually prized most as worth a life after death, are those things least liable and least truly worthy to outlive the body. It is individuality, peculiarity, or the specific difference of self that is hugged with an exaggerated care and fervor. But all spiritual progress is progress out of individualism and peculiarity and difference. Reaching and gaining true personality is leaving and losing true individuality. As we approximate perfection we become nearer alike, and the ideal of all perfection of character is that of God, all imperfections of finiteness having been left and the equilibrium of all attributes attained. The belief in any specific difference of essential being is largely begotten and nourished by those differences of organization and corporeality which are constituted by the accidents of incarnation. It is almost impossible to think of them as not rooted solely in matter. As we travel inward along any or all lines of sensation, each organ referring elsewhere for its *raison d'être*, and as we follow out the reference until is reached the last secret chamber beyond which there is no reference, we find that we have long since left behind us most that renders us recognizable, or individually peculiar. We have thus attained a purity and perfection of our deep inner being that is nearly or quite identical with that of the essential being of everything else that lives, and like that of pure spirit before clothing itself with individuality and materiality.

Whether a spiritualized and noble personality can hold unified, through and beyond death, the cluster of peculiarities and attributes which, if existing, are uncaused by the accidents of organization, and that therefore do not end with disorganization, this peculiar cluster that we specifically distinguish as our friend—this neither God nor reason nor experience has certainly told us, and for this doubtful, hopeful, blessed ignorance, let us thank God, and seek not too curiously to raise the veil. Perhaps our ignorance is proof of the comparative unimportance of the question. As the highest souls value the distinct and separate individuality ever less, and always seek the perfection of non-individual personality, so our means of the attainment of genuine immortality is in that power of spirit over time whereby the future is always present and mortal life is merged into the immortal before death comes. Such souls have long known that materiality or corporeality is the scaffolding about the steeple of life (the tower supporting the scaffold as the spirit does the body), and when the spire is completed, the workmen and death remove the hiding framework of both buildings to allow the self-sustaining column to stand forth in naked beauty and aspiring strength. And as in everything else, the beauty and strength, both of spire and of spirit, will approximate the ideal exactly in proportion as each has escaped the peculiarities and defects of individualization, and has realized the unique purity and perfection of the divine Architect.

CHAPTER XIII.

ETHICS.

THE intimate union, and what is more, the necessity of a union of determinism and the various forms of materialism that are popular, as Spencerianism, agnosticism, or atheism is of course admitted. In a world of pure mechanics there is no possibility of ethics, and no use in talking about ethics. All things that happen are necessarily so, flow rigidly from pre-existent causes, and all attempt at influencing the course of self or of the future is the silliest nonsense. The common degraded form of "science" and the "logic" of popular "evolutionism" seek to make morality a mere part of the great physical grind of the great blind machine. All such logicians make the everlasting mistake of confounding *post hoc* and *propter hoc*. Because morality and the moral sense *followed*, it does not result that it is *derived from* the material used in the preceding condition. The truth is that some aspects and ideals of ethics could find no room or function in the early stages of the development-process. All early stages were but preparations for it. Moreover, what is evolved must have been involved. Properly speaking "evolution" is development, emergence, or the unrolling and outcoming of involved plantings. It is a sad commentary on our logical acumen that the crude atheists and materialists have been allowed to grab "evolution" and "science" as if these were their sole property, when in fact they have no earthly right to the "stolen goods." True Darwinism leads to other conclusions. As Wallace says: "The Darwinian theory even when carried out to its extreme logical conclusion, does not only not oppose, but lends a decided support to, a belief in the spiritual nature of man. It shows us how man's body may have been developed from that of a lower animal, under the law of natural selection, but it also teaches us that we possess intellectual and moral faculties which could not have been so developed but must have had another origin, and for this origin we can find an adequate cause only in the unseen universe of spirit."

All questions as to the right use of life resolve themselves finally into two: questions as to being or doing, the first relating to character, motives, and ideals, and the second relating to conduct, use of power, and the like. In a final analysis there is, of course, but one, because action will more or less perfectly, and sooner or later, depend upon character. Ideals are finally worked out into practice, as the theoretical atheism of one century is followed by the practical selfishness and immorality of the next. What a parent secretly desired but showed little evidence of, breaks in the child into most literal actuality. For this very reason the distinction, while not obtaining in a logical sense, has all the more practical validity and use. The present-day disgraceful struggles between "capital and labor" are not the last results we shall know of the scientific doctrine of the "struggle for existence," and the *bellum omnium contra omnes* so zealously and exaggeratedly taught us for twenty years.

Systems of ethics are as numerous as systems of religion,—the latter, indeed, gaining most of their power and authority from their ethical quality or teaching. The very multiplicity of the systems, and their diversity one from another, together with the lack of any unanimity of opinion or practice on the part of mankind as regards highly important ethical questions, all show the want of a perfectly satisfactory root-principle and accepted standard, whence all ethical motives shall flow, and whereby all doubt may be satisfactorily decided. All abstract rules, commands, or laws, as *e. g.*, those of the ten commandments, besides being negative in character, and so supplying no incentive or motive, are by common law and common consent disobeyed every day,—thus showing that a deeper unenunciated principle gives them authority or denies them validity, by a right that is not visible in the statement itself. *Thou shalt not kill* is legally contradicted in the execution of criminals, and the reverse of *thou shalt not covet* is the motive of all tradesmen. Examples are just as unsatisfactory; to follow that of Jesus would end in ludicrous instead of genuine tragedy, and if all Christians were as sincere as Paul no civilization would have arisen. All utilitarian systems are concerned only with doing, not being, and would reduce life to as inartistic and soul-deadening a process as that of a Shaker village. The evident truth is that in the death of religion the world is drifting with the winds and currents of chance and passion, the self-interest of one being limited by the self-interest of the others, legal practice and theory being a sad jumble of historical precedent and experience, as antiquated as its phraseology, and crudely dovetailed into new conditions, problems, and necessities, that require a broader ethic and a more certain perception of the object of life.

The fundamental fault of all ethical systems has been

that to the common work-a-day mind they have offered no convincing *raison d'être* for themselves and their commands, no self-evident reason and appeal that could reach the simple understanding, no positive incentive that could move the sluggish or selfish heart. Translated by the common thought, morality is a good thing for self to preach, and for others to practise, and much of the ingenuity of mankind has been expended in encouraging ethics in the practising simpletons, and in discreetly eating the excellent chestnuts with unburned fingers. Everybody is encouraged to be self-sacrificing—with a wink to the wise. Moreover, God or conscience has seemed to command morality, without in the least practising it Himself, the morality of the general biological world being apparently quite different from that of the religious and didactic teaching of the Church elders.

Buddha's religion of kindness and sympathy takes in the animal world, but leaves out the vegetable kingdom, whilst Jesus has no word implying the inclusion of either. In the eleventh commandment, however, there is the wondrous law of love to human brethren, which had it been broader in application, and mentalized, not simply left as a beautiful but impracticable emotion, or desire for an emotion, might have served at least as an ideal dream of life.

Sentiment, however, has outrun the logical and intellectual statement or systematization, and in the modern sympathy with animals (a subdominant in much biological study) it has unconsciously (how often He does things so!) reached forward toward a truth whose significance has already made itself felt in legal statute and social custom. The societies for the prevention of cruelty to animals spontaneously springing up over the civilized world are simply founded in obedience to an unconscious recognition of the essential unity of all living beings. Whilst seemingly negative in function, the emotion is positive in character. But no formularization of the principle, psychical or biological, underlying and stimulating the emotion, has been expressed or recognized, and not even the most tender-hearted has dreamed of suggesting an extension of its application to the plant world. But from a financial and civilizational point of view it is infinitely more important that Societies for the Prevention of Cruelty to Trees should be formed than those in reference to animals.

As a general rule, needing some modification and explanation, all rules and ideals of morality, both of action or being, are perfectly included in the simple injunction to imitate God, to do what God is doing, and to be what He is. The idea of taking God as a model of action, and especially of character, may seem to some impious, and perhaps it will seem so to them whose Divinity commanded, "Be ye therefore perfect as your Father in heaven is perfect." To adopt such an ethical principle of imitation of God requires, of course, first the most absolute conviction of the existence of God, together with what kind of a being He is, and secondly, there will be required a sympathetically intellectual science of what God is doing or aiming to do in the world. The excellence and perfection of this ethical principle is that it is properly essentially positive, but permits the easy negative. It is applicable to the lowest immoralities, such as murder, or adultery, whilst it also includes the most exalted ideals in its comprehensiveness. It puts both a principle and an example before us. It is adapted to the biological process, extending a principle of conduct over the neglected animal, vegetable, and even physical world, and making morality an affair not of sociology alone, but of physics, physiology, and nutrition. It is progressive, and it supplies a flexible but not breakable principle that may command one thing to-day or in this instance, or the reverse to-morrow or in another instance. Lastly, it makes moral action partially dependent upon intellectual progress and clearness, since to formulate the entire motive and plan of God's action in the world is distinctly though not solely an intellectual task, and in adapting the principle to the strange diversities and conditions of life, room is left for the educational play of reason and enlightened casuistry. Thus are we not bound either by a sorry jumble of inflexible rules and objective commands, or by a subjective blind tyrant-instinct to do the right, whilst we are left without a ray of light to know what the right is. (It is useless to deny that the greatest crimes against God and man—*e. g.*, the Inquisition, wars, and murders—have been done with the sincerest conscientiousness.)

An ethical principle should be positive and exceptionless in character, the negative and qualifying aspects being supplied by the application, the circumstance, and the light-giving intellect. Biologos authoritatively commands, *Increase, multiply!*—but experience and reason dictate monogamic marriage, together with many other qualifications of the same not yet legally approved or disapproved. The common intelligence and state of civilization will also have to extend the application both of positive and negative command over the individual, because the common is greater than the individual experience, the common good and safety greater than the exceptional good and safety, and the common wisdom is greater than that of an average individual. The individual must square himself with, if necessary be sacrificed to, the community, in all the lower and fundamental aspects of ethics, because God's incarnation-process is dependent upon "preservation of the type."

If the fundamental rule of all ethics is that of doing what God is doing and of approximating Him in character, several things will follow, but notably this: they who do not see and acknowledge the divine intelligence at work all about and in themselves must, by the fact itself, be limited to the rule of God (for rule He will and must) by the same methods He uses in the animal and vegetable world —that is, by instincts, the conflict of forces, the blind working to ends unknown and devious, the control by the lower factors, such as nutrition and sensualism, and by the thousand indirect and subtle means that God has of making His direct cell-controlled agencies bring about His indirect larger and spiritual purposes. According to the older theological phraseology, or an extension of it, they must be held in obedience " to the law." Acknowledging and choosing God as model is the advance into a new kingdom of love, and at least when general, the fact itself will constitute the greatest advance, biologically speaking, that has so far been made. Because a thousand things show the manifest awaiting for help on His part, the great salvage of wasted force, the end of war and of luxury, the unprecedented gain from order and systematization, that is to come when the blind machine itself turns and becomes the conscious and sympathetic helper. But the biological or incarnational aspect is not the only, perhaps not the chief one: there remains the profound change in character that at once follows this free choice of the divine ideal. The human animal, to whom has already in great part been given the control of the results of the incarnation-process,—the human animal, still unconsecrate deputy, is by this sympathetic choice lifted into true personality and co-partnership with God.

It seems to me that only in this way, and speedily in this way, will be brought about the genuine Kingdom of Heaven on Earth. This beautiful dream of Christ and of many of His followers rightly consisted primarily in a "change of heart," a recognition of the Father on the part of the child, a self-consecration to that Father, and, so far as conceivable, a conscious aiding in His work. But the dream was necessarily dream-like and resultless because of two or three unavoidably conditioning characteristics: it was not only inceptionally, but it was throughout a matter of sentiment or emotion, in which the intellect had no share; it ignored the entire subhuman part of the incarnation-process—that that we term science or biology; and even of this human part it would utilize only a fraction of the agencies operative in society. All those of political economy, mechanics, government, education, art, and the like were utterly ignored. It lacked direction somewhither, and also the very mechanics of progression. There was a profound objectlessness of purpose that, like the typical Christian movements of the Crusades, led nowhither with fiery, useless, and tragic zeal.

This subjection of the human animal (or at best the instructed deputy) by the indirect agencies God is compelled to use with His children that remain obstinate and unrecognizant of the Father's love and work,—this divine compulsion will in the future be progressively supplemented by an unconscious subjection to the authority of those human sharers of God's kingdom who willingly consent to use their power for His purposes. Invisible forces are distinctly at work to turn the world of selfishness to unselfish and public purposes, and to make the most disobedient and proudest self-seeker the servant of the most obedient and meekest God-seeker. A thousand examples quickly appear before the thought: conscience-stung plebeians devoting ill-gotten wealth to public uses, founding libraries, universities, or observatories; selfish and narrow-minded men gaining legislative and executive power for selfish ends, but forced to forego those ends and to use the power for purposes better than they know by men and influences they can only secretly hate, and can never understand; a tory or aristocratic party committing suicide by "dishing the liberals,"—enacting measures of democratic tendency; money-seekers inventing and using machines and labor-saving devices that help humanity, or building railroads and telegraphs for purposes they as little understand as they do the nature of nerve-currents. I know many instances of bad men curbed, held in place and utilized for larger and better purposes by good men they hate, the good men never using their secret power selfishly. There is a distinct modern growth of this fact of holding lower personalities and powers to a larger and impersonal use, and of making the private path of selfishness lead to another's garden, or to the public highway itself. This delegation to the unselfish of God's own secret and subtle methods of rule is, of course, aided by the ready perception on the part of those who understand genuine values, that most of the ideals of men do not lead to genuine power or happiness, and that consequently what is usually called happiness is but a mockery and a delusion with which men are enticed to throw the neglected but real results of their lives into the lap of humanity. Thus it comes about that minds that comprehend real values enjoy the parks of the rich more than the owners, do not in the least envy the owners their sad pleasures, read the books their grudging or tainted generosity has supplied, and in every act utilize the labors of the poor servants who were both incapable and scornful of a proper utilization of the civilization they unwittingly provided. This direction and control of the world by unknown minds for unknown or public purposes is destined to great extensions. God is only waiting for the sympathetic and intelligent helper. The Napoleons are to be fewer, the Lincolns more numerous, in the future. The difficulty is to unite feeling and intellect, to emotionalize reason, and religionize science, and when this is done, power awaits the user. The Kingdom of Heaven is indeed at hand, but the Kingdom of Heaven will be an absolute democracy, temporarily intermediated by an aristocracy of the humble lords of life.

The extension and perfection of healthy life over the globe is undoubtedly the plainest aim, and the most primary work of Biologos. Whatever aids in that is right and whatever opposes it is wrong. To gain a foothold in matter, to make this foothold secure, to extend and ever perfect the simple fact of life, is the fundamental and, up to now, much the largest part of His concern. The incarnation-process has been attended with such tremendous labor, difficulties, and dangers, that the first of all sins is murder, and the first of all commands is, *Thou shalt not kill.*

In the working out of this incarnation-problem and task, Biologos has had Himself to resort to a thousand seeming contradictions of the law of life, because the lesser has often had to be sacrificed to the greater. He has had to kill in order that universal death might not follow; He has had to sacrifice the single or the few weak, old, diseased, maimed, or defective, in order that all might not be sacrificed: and He has had to make the struggle for existence a great means of progress in animal mechanism and diversity, (as all Darwinians are very fond of telling us), death itself being one of the conditions of greater and larger life. In the animal world or in the human-animal world, if polygamy, sexual war to the death, or whatever other strange mechanism of reproduction best served the temporary exigency, and best insured the stability and extension of the type, and hence of Life's incarnation-work,—this was grasped and resolutely carried through. The fact was all important; Life's foothold once secure, correction and perfection could follow. The entire biological process offers million-fold examples of a compelled doing evil that good might come, or of sacrificing an ideal morality to a present necessity. God's transcendent freedom could see through means to ends, and when our motive is as pure, and our reason anything like as pure as His, we may safely adopt even His use of freedom, or imitate His disposal of means.

But it must be remembered that His necessity is not so stringent now as in the past, and that moreover His necessity is not ours. In human life the nutritional problem is now so far solved that only in rare and exceptional cases is killing a necessary means to the progress of human life. But when killing is such a means, we ourselves do not hesitate to do it, while as regards the animal world we kill without a thought of sin, and we would laugh at the charge of sin in unnecessarily destroying a plant.

Finally I must again emphasize the easily forgotten truth that we are all His work, nay, that we are He. Our highest ideals of the right are of His planting, and in His soil that has been manured with a thousand deaths. This and many such things are quite as much divinely-caused facts as two seal-bulls fighting for possession of the harem,—nay, they are more a fact, because the seal-war is a temporary expedient, whilst in the pure human mind the ideal is the true coming to light of the delayed fact, the blooming of the plant of the long-cherished exotic seed to which all else was preparatory, the plant that could not have come to fruition in the savage conditions of a more violent age and clime.

The taking up of God's work in the world is therefore the true ethical standard of human conduct, and the furtherance or extension of healthy life is the chief and fundamental part of His work. But I wish to lay especial stress on the truth that whatever difficulties may lie in the way of making the principle the basis of a systematic and theoretic ethic, there are few difficulties in the way of an enlightened mind making it both the motive and standard of a practical and every-day ethic. It is comprehensive in its largeness, whether of positive energizing motive or of limiting and qualifying application. Critic folk and lordly logic-slaves may easily rail at it, but if a clear heart and clear mind will seek to carry it out in practice it will be astonishing how ethical doubt and perplexity disappear and subjective difficulties are merged into objective necessities and utilities. The mystery and vagueness of a blind sentiment or propulsion of conscience are brought out of the darkness of individual or racial experience, education, inheritance, and custom, and set in the sun's light of reason and intellect. Here they are adjudged and rated by a divine and cosmic standard. We cannot go wrong if we adopt God's rule of right as our own, and there can be no doubt as to the value of the progressive incarnation-process in God's mind.

It should also be carefully observed that however strange the principle appear, and however unacknowledged it has been, it has in most literal truth been the unconscious basis and standard of all that was valid and good in ethical systems, judicial rulings, and in religious and didactic teaching. It would not be difficult to show that the neglected laws, reversed rulings, ignored precedents, and rescinded enactments that experience has taught men thus to treat, or those that have brought ruin and wrong in their fulfilling, have been bad solely and accurately because they were not consistent with or were opposed to the progress of the incarnation-process, *i. e.*, to the extension and perfection of life in the world. Wherever the most deep-rooted conviction of right, the most accepted truths of religion, or the most profoundly ethical of principles have clearly contradicted or placed obstacles in the way of the incarnation-process, as has often happened, they have been swept aside as cobwebs, ridden over even with enthusiasm, and ruthlessly crushed beneath the unfaltering hoofs of this spurred-horse of necessity. Strange as it may seem, too, with sudden ability men have twisted their prejudices into conformity with the necessity, and have even made the unwilling priests of peace bless the very battle-flags of war. All history is illustrative of this fact. It not seldom offers a ludicrously sad spectacle of human duplicity and self-deception, as, *e.g.*, when a people and church profoundly believing in the sanctity of marriage and hating adultery authorize divorces, and encourage any form of legal adultery, rather than that lack of an heir to the throne should make civil war probable. The most heinous sins of one's own party are sniffed over and ignored, whilst those of the opponent are most frightful to behold. Each of two Christian armies on the eve of battle, with equal sincerity prays the God of Peace to give self the victory, and to the enemy death.

Non-recognition of the difficulties and dangers of the incarnation-process, and that the attainment of the end has temporarily necessitated the adoption of any means thereto best suited, have been at the root of much atheistic or sensualistic misreading of biological fact and history. Men have either exultantly or with deep pain seen a supposably profound antagonism between biological and divine morality. It seemed that there was an absolute contradiction between the God of evolution and the God of conscience. Hence irony and pessimism, a disobedience both of the Divinity in the world and of the Divinity in the breast, or a shutting the eyes to one set of facts, a scorn of conscience, or a scoffing at science,—and in either case a sad narrowing of character and denial of intellect.

I wish also to note that the acceptance of this as the fundamental ethical principle of human conduct tremendously and beautifully broadens our ideal and possible attainment of freedom. The imitation of God is necessarily coupled with the distinct sharing of His kingdom, the according to us by Him of true autonomy, and, so fast as possible, the extension to us of genuinely creative or divine power. In any other conception of the divine character, and of the method of His government of the world, there is in the breast a consciousness more or less clear and more or less hated, of the inescapable tyranny of His all-comprising personality. The logic was irresistible that, consistent with the premises, no free subordinate personality was possible, and that all other personalities than His were necessarily servant-like. We all sympathize with the child that perpetually wished there were some room in the house where God could n't find or see her.

According to the idea I have tried to bring out, we are in most literal fact children of God—that is, in a last analysis God Himself, devoted to a special task, and temporarily renouncing those divine attributes not needed in His work. Such a perfection of result in the special task has now been reached that there is taking place a resumption of the attributes of the pre-incarnate divinity. This resumption of the Godhood is the process of the spiritualization of character of civilized and educated mankind upon which we are entering, and this I prophesy will be somewhat sudden. At the end of the evolution-drama man, the hero of the play, removes the tragic mask, and lo, it is the benignant and smiling face of God behind! We are on the eve of a wondrous and glorious perfection, a true divinization of the human mind, whose explanation and characteristic on the divine side is this, that having perfected the nutritional mechanism that was the fundamental problem in incarnation, having brought his incarnations to a power of control of molar motion, having intellectualized that control, it is now to be spiritualized and devoted to genuine use. Useless dead worlds and aimless mechanic forces are to be utilized and devoted to the purposes of spirit. Having so far reached the object of incarnation, absorption in the work is not so profound, and there must now be an enlarging of the divine view, a reclothing with the divine freedom. To us this is simply a breaking-through of the divine consciousness into the human, or rather a recognition on the part of humanity of its genuinely divine nature, the acceptance of offered sonship and proffered kingdom.

In a systematic ethic our duties or relations will phase themselves somewhat differently in six groupings of facts: 1. Those to the inorganic world. 2. To the plant world.

3. To animals. 4. To our fellow-men. 5. To self. 6. To God. Taking God as our model, let us briefly indicate some of the duties and aspects of our conduct toward each of these classes.

The purely inorganic world constitutes the material which the divine Architect uses in constructing his House of Life. A human mechanic is not successful if he does not understand every quality and power and "law" of the material he uses. The study of physics (including chemistry), therefore, lies at the foundation of all true education, understanding, and use of the world. The inherent modes of action of every element and of all non-living combinations of elements, under all conditions of temperature, pressure, circumstance, and the like, should be the primary part in any advanced course of study, and especially in the mechanic arts and sciences. The clear distinction between the use of atomic or molecular motion, and that of molar motion, lies at the beginning of a comprehension of the mechanism of incarnation, and of the use of mechanical force that Biologos reaches by the combined cell-activity of muscular function. The progressive study of these infinitesimal forces is already reaching toward a science of cytology that shall reveal to us the secret of the mechanism of His entrance into and use of matter. Aided by perfected thermo-chemistry and physiological chemistry, the revelation at a not far-distant date will be complete. It is our duty, therefore, to know and understand the inorganic universe of the infinitely great and extended, in order to orient ourselves and to judge of the place we hold in the sidereal and solar system; also of the infinitely small, in order to understand what He first must have known so well, that infinite detail of work which has constituted the means and mechanism of His incarnation-process. But He not only knows: He knows in order to use; and, as we are certain, He uses for purposes of spirit, and to extend the reign and power of Life. So must our control of the inorganic world be devoted to a like end, not used only for selfish purposes, not for the good of our own life alone, but for the good of Life. Every victory of knowledge, every invention, every device for increasing our domination over the dead forces of the world, should be turned at once to the use and purpose of all life,—the winning to Life's control of hitherto unutilized forces, and that spiritualization of the biologic process which is clearly the aim of the biologic Father. The utilization of the molar motion gained by muscles, and especially by mechanical inventions, is the pronounced characteristic of this age, and this utilization constitutes the systematization and firm safety of the nutritional process by which the incarnation-process is carried on. The overcoming of past dangers and difficulties will now enable Biologos almost suddenly to bring the Kingdom of Heaven on earth. Patents on inventions should therefore be short-lived, and on many things patent-right should not be allowed at all. The utilization, for the purposes of Life, of the now-wasted forces of the tides, water-falls, winds, sun's-heat, electricity, and the like, should come about as quickly as possible. The discoveries in "pure" or theoretical science, upon which at last most practical inventions depend, should be furthered by governmental and common encouragement, the results becoming the property of the people. Those who simply make application of the great generalizations to practical use, should not be given too great or too long a monopoly. Profits above a certain percentage may rightfully be restricted. Humanity's rights are pre-eminent in all things, and especially in all such things as these. A great reform in government should be that of gradually turning it from what at present constitutes a large part of its function, the means of aristocratic, plutocratic, or class aggrandizement, and making it a means of furthering original research in physics and biology.

Our relation to the world of vegetation is plainly different in one important respect from our relation to the inorganic world: we have to deal with Life, and even in its lowliest aspect this commands our respectful love and reverence. The same spirit that lives in us lives also in leaf and tree. The "angry tree" of Nevada, if disturbed, shows evident vexation, "ruffling up its leaves like the hair on an angry cat, and giving forth a sickening odor." The "stinging tree" of Australia also emits a disagreeable odor, and causes excruciating pain, even for months, in any one who gets "stung" by it. "Dogs and horses have sometimes to be shot" when thus wounded. A strange protective device is that of the phosphorescence of the "witch tree." In Taylor's *The Sagacity and Morality of Plants*, there are described a thousand delightful proofs of the mentality of plants. The plant expresses its life as best it can with the means at command, ever straining toward more perfect command of means. The leaves of Dionœa require summations of stimuli exactly like our own muscles, to effect contraction. They are likewise fatigued by repetitive stimulation, just as is muscle-substance. The only difference is in sensitiveness—*i.e.*, a more perfectly acting mechanism. There is no difference in kind, only in degree. Under a pressure of three hundred to six hundred atmospheres yeast and muscle tissue react in an almost identical manner. The law that, where food is denied and the struggle for life is hard, a preponderance of males is produced, holds good in both the plant and animal world. One spirit or life exists behind all living things; all orders or types are bound together in a beautiful interdependence. The entire scheme of incarnation depends upon the proper adjustment and co-ordination of all the parts. The working of the engine depends upon the integrity and action of the smallest part.

Our use of this plant-world must, therefore, be not solely as dead material, to which we owe no duty or respect for its own sake. That it is for our conditional use is plainly shown by the fact that it is necessary to animal existence, its complex cell being the animal's necessary food. The stored heat and energy of the highest cells perfected by the plant are the raw material and means of that more bewilderingly complex cell-life that constitutes the physical mechanism of the animal economy.

It is therefore evident that man is interested in the vegetable kingdom not for his own personal sake alone, but the fate of all humanity and of all animal life depending upon the lower order, it becomes his interest and duty to preserve and care for the lower in its entirety, as he would the foundations of his house, and the integrity of his government. The clearest evidence of the common duty of humanity toward the general plant world consists in the sin of deforestation. That exquisite balance established by Biologos between the tree and the still humbler orders of plants, and between the animal world and the vegetable, is evidenced by the deserts of the world, with changes in temperature, rainfall, and other climatic conditions that endanger the very existence of life on the globe,—all of which has followed man's ruthless destruction of forests. Professor Buckland says there is no such thing as an original desert on earth, all deserts being due to man's destruction of trees. The whole question is of vastly more importance than any that has occupied the attention of any congress or parliament in latter years. The repeated warning still falls on deaf ears. Every government should at once make it a crime to build a wooden-walled house. How far wood may be used in the smaller parts and finishings of a house should depend upon many circumstances, but sharp restriction must soon be exercised even here. Life's labor is great enough to fashion her necessary and lowliest structures; to sacrifice them for purposes better served by a world full of stone and minerals ever at hand, is a shameless squandering of precious wealth. But when the sacrifice invites a second sacrifice to fire,* and when it endangers agriculture, even civilized life, it becomes enormously pernicious. What more consummate idiocy than to encourage our own country's deforestation by a tax-prevention of foolish outsiders from sending us their murdered trees.

What a stupid shortsightedness is also that of our neglect of plant diseases. The loss from "rust" in wheat is each year a hundred times as much as has ever been spent in all the experiments and study to prevent it. The loss by "smut" in corn and oats is as much as ten per cent. some years; millions of dollars are thus wasted, and not a dollar spent to learn how to prevent it.

But it is not enough to emphasize the negative aspect. Trees must be encouraged, sympathized with, and helped to realize their desires. Very few trees there are out of

---

* The average loss by fire in the United States is estimated to be about $500,000 a day, i. e., half a million people are kept at work all the time to recoup the nation for its fire-losses.

millions that do not bear as many evidences of their struggle against difficulty and denied nutrition as of victory in life.* Very few there are that do not show desire and beauty hindered and checked, pathetic exhibitions of struggle against adversity † and of frustrated aspiration. If with all our spiritual and mental life we were to become a tree, we should act and show our life precisely as a tree now does to us. One tree that has realized itself is worth a journey of a thousand miles to see. God loves His dear trees, and we should love them. They can teach us many lessons that we need, as they are a result of the loving labor of God, and of His delighted success in a beautiful method of His self-incarnation. Their work in transforming the earth into a home is so exquisite, and their important influence on rainfall, climate, water navigation, and agriculture is so indirect, that their own life is allowed a self-expression and a perfection that is denied to organisms whose work is more immediately or quickly utilitarian. Nothing attests the greatness of the character of Moses better than his order to his army that in besieging a city no tree should be destroyed. It is

* The beautiful observations of my friend, Dr. Wilson (*Contributions from the Botanical Laboratory of the University of Pennsylvania*, 1892) show how intelligently responsive plants are, and how they adapt themselves to the varying conditions of sunshine, darkness, cold, and heat ; we see plainly that the plant has its own troubles, trials, needs, habits, and all, like ourselves.

† As, *e. g.*, in the war with cold in cold climates, and with heat in the torrid zone. The action of bark as a non-conductor is remarkable. Twigs are not frozen in our winters, and in summer the temperature does not increase with that of the surrounding air. Hooker mentions a fruit grown by the Ganges in a soil having a temperature of 90° to 104° F., but that of the juice is only 72° F. The action of trees rooted by crumbling banks, their twists, turns, and desperate struggles to keep their footing and uprightness, properly arouse in us pathetic and sympathetic emotions.

strange how slow people are to recognize even the commercial value of scientific forestry. How many men there are that will work life out in cruder ways to make their land valuable when a few planted trees would in time and with no expense soon double the land-value. But we are in such haste, we cannot wait for the tree to grow; its lesson of slow and patient growth, reaching through several human generations, has not been learned by the restless human heart, that grasps at green fruit and creates artificial desires.

Every improvement in agriculture, and especially the new and promising art of scientific agriculture, looks toward both a happier farmer, the basis of all right civilization, and toward a healthier and happier city. And this not alone through an increased and more perfect food-supply; besides the raw material of food furnished the city, the country must also feed it with healthy and moral men and women. The decay of rural New England is a sad and awful spectacle. To the sneer of his big-ideaed brother returning from the West, the Yankee farmer proudly answered, "We plant school-houses and raise men." We have not yet more than begun to feel the curse that is coming from our tariff sin of grinding up the farmer in the city mills of the manufacturer. God, it is said also has His mills, and they proverbially grind slowly, but most exceeding fine.

One of the excellencies of agricultural and rural life is the practical example of the fitting union of animal and vegetable life. The trees along watercourses protect the banks; along roads they shade, protect, and beautify them, and the cattle find resting-places for rumination beneath them. To make two blades of good grass grow where before but one grew, or where there were only weeds; to give healthy pasturage to healthy animals; to perfect dairy science and art and thus provide better means for Biologos to fashion man's best food; to increase the grain, vegetable, and fruit supply,—to do all these and a thousand such things is to imitate and to help God. And He knows how to reward those who help and practically love Him. Stay in the country, young man, but enrich your life with all the results of city intellect; the railroads and books now give you the opportunity to do this.

If the plant is worthy of love and respect for its own sake how much more so is the animal; and yet it is only with difficulty that our civilization is beginning to recognize the rights of animals and our duties to them. Every one with refined sensibilities of justice, and recognizing the unity of animal and human life, shudders at the butcher shops, and the slaughtering of animals for human food. Should one be a vegetarian? I think not. But if the most complete cell-life and the highest development of human life, as seems probable, need animal food, they do not need it to the frightful extent now indulged in, and the procuring it should not destroy any type or species of animal. It should also be carried out without suffering. Pain must be abolished; our own nervous system demands that, even if it were not evident that God gave nervous pain to guard against suffering and danger, not to continue itself. Any "sport" that requires the useless suffering of our lower brothers must be stopped by law, and with the sharpest penalties. Hudson's thrilling and mournful words so well express my own feeling that I again quote him:

"The rhea's fleetness can no longer avail him. He may scorn the horse and his rider, what time he lifts himself up, but the cowardly murderous methods of science, and a systematic war of extermination, have left him no chance. And with the rhea go the flamingo, antique and splendid; and the swans in their bridal plumage; and the rufus tinamon—sweet and mournful melodist of the eventide; and the noble crested screamer, that clarion-voiced watch-bird of the night in the wilderness. These and the other large avians, together with the finest of the mammalians, will shortly be lost to the pampas, utterly as the great bustard is to England, and as the wild turkey and bison will shortly be lost to North America. What a wail there would be in the world if a sudden destruction were to fall on the accumulated art-treasure of the National Gallery, and the marbles in the British Museum, and the contents of the King's Library—the old prints and medieval illuminations. And these are only the work of human hands and brains—impressions of individual genius on perishable material, immortal only in the sense that the silken cocoon of the dead moth is so, because they continue to exist and shine when the artist's hands and brain are dust: and man has the long day of life before him in which to do again things like these, and better than these if there is any truth in evolution. But the forms of life in the higher vertebrate classes are Nature's most perfect work; and the life of even a single species is of incalculably greater value to mankind, for what it teaches and would continue to teach, than all the chiselled marbles and painted canvases the world contains; though doubtless there are many persons who are devoted to art, but blind to some things greater than art, who will set me down as a Philistine for so saying. And, above all others, we should protect and hold sacred those types, Nature's masterpieces, which are first singled out for destruction, on account of their rarity, size, or splendor, and that false, detestable glory which is accorded to their most successful slayers. In ancient times the spirit of life shone brightest in these; and when others that shared the earth with them were taken by death, they were left, being more worthy of perpetuation. Like immortal flowers they have drifted down to us on the ocean of time, and their strangeness and beauty bring to our imaginations a dream and a picture of that unknown world, immeasurably far removed, where man was not : and when they perish, something of gladness goes out from Nature, and the sunshine loses something of its brightness. Nor does their loss affect us and our times only. The species now being exterminated not only in South America but everywhere on the globe, are, so far as we know, untouched by decadence. They are links in a chain, and branches on the tree of life, with their roots in a past inconceivably remote ; and but for our action they would continue to flourish, reaching outward to an equally distant future, blossoming into higher and more beautiful forms, and gladdening innumerable generations of our descendants. But we think nothing of all this ; we must give full scope for our passion of taking life, though by so doing we 'ruin the great work of time'; not in the sense in which the poet used these words, but in one truer and wider and infinitely sadder. Only when this sporting rage has spent itself, when there are no longer any animals of the larger kinds remaining, the loss we are now inflicting on this our heritage, in which we have a life-interest only, will be rightly appreciated. It is hardly to be supposed or hoped that posterity will feel satisfied with our monographs of extinct species, and the few crumbling bones and faded feathers, which may possibly survive half a dozen centuries in some happily placed museum. On the contrary such dreary mementoes will only serve to remind them of their loss ; and if they remember us at all it will only be to hate our memory and our age—this enlightened, scientific, humanitarian age, which should have for a motto, 'Let us slay all noble and beautiful things, for to-morrow we die.'"

The death of those most exquisite of God's creations, heaven's own birds, for purposes of ornamentation of our female savages, is a disgrace to our humanity. It is grievously sad to think it has gone on so long. Let a woman so cruel as to wear a dead bird be shunned as a murderess. She should not be spoken to by good people.

The best of stabling, food, water, and care is certainly due the animals that willingly work for us and give their lives to us. If a very limited use of vivisection experiment is necessary for scientific and medical progress, it must be regulated by law, carried out with jealous guarding against excess and against suffering, and the maimed animals painlessly killed when the experiment is complete. The practice carried on by conceited jackanapes to prove over again already ascertained results, to minister to egotism, for didactic purposes,—these are not necessary and must be forbidden. Every kind person should join and help carry on the societies for the prevention of cruelty to animals.

Every animal body is a product of marvellous skill, that can always teach us lessons of interest and use, and every animal soul is capable of education, sympathy, and progress. Fashioned so exquisitely by Him who fashioned us, His life in every cell, His spirit forming the essence of the personality of each, it becomes us to call forth the latent capacities of animals, and cover them with our care. To our loving sympathy there is always on their part an eager response, and many a man has been taught deep lessons of soul-life by them. There is yearning in their eyes for our teaching, invitation in their very artlessness, and the readiness with which they catch soul of us tells how willingly their life will rise to meet yet greater opportunity when it shall be offered.

Unto a few the exigencies, difficulties, and accidents of the incarnation-process have given tremendous influence over their fellow-men, and to every one some measure of power. If the humblest will but think carefully he will recognize how great and frequent has been his influence either directly or indirectly upon others ; and by reflecting more seriously he will see how much greater it might have been, and how in future it may become much more marked. The responsibility thus given to us is negatively or positively ours, and whether we will or not, and our use of it becomes the chief opportunity and duty of our lives. In these matters sins of omission are as great sins as those of commission.

The difficulty of deciding how to use the power given us, or that might be ours if we but grasped it, the vagueness of ethical ideals and the impossibility of living up to them, have served to make men withdraw into themselves, throw away responsibility and opportunity, and drift with the chance current of tendency and *Zeitgeist*. The aspirations and ambitions that swell in our breasts when we are truest, tell us what is the aim of our essential life, what God would like to do with and through us. The impulsion of humanity toward progressively better life, and the inherent tendency to progress observable in civilization, both show us the simple ready duty ever at hand : to complete the process of humanization ; to aid in the progress of civilization ; to help men, through whom God is working, to realize the ultimate ends of the incarnation-process. If this process have any meaning and final cause, it must be outworked in and through humanity. Hence of old the chief duty of man has been held to be toward men, and however exaggeratedly this aspect has been emphasized to the sad exclusion of other duties, it must still remain the chief of all. And this is true principally because the ability and opportunity are greater in this direction. Every day we act upon and with others, and the great temptation and sin of each day is to use or to seek to use others for our individual life's good, instead of for the good of all life. It has always, and by all teachers of morality, been clearly seen that selfishness is the predominant sin. Why it was so was not clear, since self-love and self-preservation is the almost invariable rule of all living things.

It seems to me that this profound contradiction of the altruistic conscience and the egoistic necessity, the opposition of ethics and biology, can only be explained by the philosophy of life that I here suggest: The egoism is the result of the extreme or hypertrophy of individuation, which under the circumstances was a necessary part of the incarnation-process; and the demand by the conscience for unselfishness is the notification of God that that extreme may and must be merged into the higher truth of the good of all life. Individuation had to be pushed to the last extreme in order to make head against and conquer the imminent dangers and catastrophes always before and close about. Hence the mechanisms, mental and neurological, of self-value and self-love had to be brought to the highest pitch of perfection in order to avert, to preserve, to live, and to outlive. Thus have arisen the sleepless, jealous, and exorbitant egoism and egotism of man. But the voice of conscience at once and early began to speak to the humanized animal that this was an exaggeration, and that this necessity and rule of life is to pass. The clear command of all high religions and true ethics, if not to reverse the rule of biological history, was at least to extend it to all our fellow-men—to love the neighbor as one loved self. Life's foothold has been secured, the incarnation-process in the human has reached such success and entrenchment that the mechanism of an enormously exaggerated self-valuation is not needed, hypertrophy must be reduced to normality, and the higher value of the whole must be recognized. The function of conscience in humanity is thus primarily clearly biological, and leads the intellect to recognize the validity of the common welfare, and also spurs the emotions to effect it. Civilization in its biological and spiritual aspects is thus at first negatively a process of reduction or subordination of the individuation mechanisms and products that have been born out of the rigors and hardships during the past incarnation-process. But positively, of course, it is infinitely more than this: It is the spiritualization of the power and energy now gained, a development of the latent and waiting ideals and objects now first possible to come to light. The rooting of the plant, its growth, security, and stored nourishment, must precede the flowering and fruitage.

One of the great reasons of the convincingness of pessimism lies in the truth, however negatively stated and neglectful of the positive truth, that unselfishness is, as it were, good selfishness, a genuine wisdom, and an excellent policy. What we gain in the long and wretched war for self-advantage does not pay for the struggle. When pursued as an end in itself, our little self is not worth the pain and strain we put into life. We, indeed, are but one tool in God's hands, that without solicitude He throws away, when we become unfit or when that part of His task is done. He seeks with us ends that are beyond any individual; His care was for a process, an object only reached through and by millions of successively appearing and disappearing individuals. The mad rush for the delusions of life, for wealth, ambition, fame, power, as all wise men know, leaves in our hands only burning ashes, and in our mouth only bitterness. Hence arise pessimism, atheism, the theoretical Buddhistic and the commonly practical Christian doctrine of renunciation; hence the frightful growth in recent years of suicide, and what is worse, the wild debauch of sensualism and luxury, in order to drown the voice of conscience.

Now to these pessimists and renunciationists there is but one answer: What you say is true,—so far as it goes. There has indeed been an over-development of the feelings and mechanisms of selfhood and of individualism. "It does not pay" to rate self so highly; there is nothing in life to warrant or to justify such an undue valuation. The individual is but a link in a chain; there is a purpose being worked out through individuals that is greater than they. There is something better and higher than individual identity, or than that persistence of partiality you desire in all the self-regarding virtues and vices. Firstly, however, be it well noted, your renunciation is a matter of selfishness. If there were more value to the delusions of life, you would not renounce. Your resignation is therefore lack of virtue; is not to be dignified by much more honorable epithets than laziness and incapacity. Secondly, your renunciation is simply negation, and you affect to ignore the fact that there is a positive object worth all struggle and energy. Your pessimism is a tricky method of self-delusion, of hiding from the mind the facts that make against you. To you it also needs be said that self-delusion also does not pay. Pessimistic pouting is a child's way of concealing the fact that the game you chose to play was not the best one, and that even that game you did not play heartily and according to the rules.

But there is a game and an object of life that does not end in tragedy or in disgust. God is playing the great game with lives, and He will tolerate no rival player except he play in the same spirit and with the same objects as He uses. His chief delight is the game of Life, and in this He loves rivals, because rivalry here means co-partnership. In this He is always on your side, since the only true opponents in this cosmic battle are Fate, and Death, and dead matter. In this great chess game selfishness is simply the ruled-out absurdity of seeking to crown your own pawn before it has honorably reached the king's row. Struggle manfully and legitimately to attain the king's row with your little pawn, and by the struggle you have learned true kingship, which is likeness to the Great King of Life; then at once crowning and power are yours! The essence of all sin is the use of the personality of others for selfish purposes, whilst the essence of God's rule and aim is to use lives for the purpose of Life. If God exist it could not be otherwise, and if you choose the reverse of His rule, disgust and pessimism are inevitable. Choose His use of life and His object as your own, and all is changed. Content and hope and sacred satisfaction are henceforth yours, and also power. Reach up and slip your hand into God's hand, with a sincere prayer to be rightly led, and lo! into your earthward hand are placed the groping fingers of Life's little children, they that are also seeking to be led to the light. All history is lurid with warning or splendid with illustration of this truth. Look among your friends and see how the mystery and the seeming illogicality of their lives, either in reward or punishment, stand revealed as the outworkings of this fundamental truth. And this, too, in a time when the vast majority of men are opposed to God and utterly oblivious of His purposes in incarnation. When a majority shall turn and adopt His purposes as their own, how manifest will then be the truth, and how sudden the now frequently delayed judgment. Love Life intelligently and heaven is yours; love your life recklessly and hell is your choosing.

It may be well to again emphasize the truth that this ideal of human duty is so all-comprehensive as to include even selfishness and unselfishness. It can explain why either may temporarily be the exclusive rule, according to the ever varying difficulties and circumstances that beset us. Indeed, unreasoned and unqualified unselfishness would almost be insanity, at least would speedily defeat its own ends and the ends of the incarnation-scheme. One has other duties, equally clear and modifying such an extreme, and which are also subordinate to the great law (a true law, a genuine superimposed necessity) of aiding the realization of God's purpose in the incarnation-process. Circumstances and conditions constantly arise in which any one duty to self, to humanity, or to lower life, must be subordinated to another duty. The exigencies of life are often excessive, and even Biologos Himself has had to sacrifice ideals, "duties," nations even, and postpone final aims to the rigorous necessities, dangers, and demands of the *fatum* of earthly accident and of difficultly-subjugated matter. Herein lie the office of casuistry, the flexibility and the plasticity of a principle that shall be adaptable to all the possible conditions of our lives, and the application of intellect to the determination of any rule of special conduct. All rules range themselves, however, with perfect obedience and clearness under the general principle, and in the most perplexing condition the most primitive intellect can still act rightly, if the heart but acknowledge the law and if the will but obey it.

Money has become such a measure of all human values that the ethics of economics is largely the ethics of social life. But financially-valuable things are worth what they are because human effort has made them so. Every dollar is but the concrete representation of human effort and thought, and everything called wealth is the product of the work, the heart-throbs, and the mental powers of others. Money is the tally-stick of muscle contractions, of heart-beats, of lives worn out. From everything purchasable with money the dead eyes of the human souls who fashioned it look out with significant demand. With every board or brick that shelters us, with every woven thread that covers us, with every morsel of food that we eat, or of pleasure that we enjoy, the shadowy ghost of humanity calls out to us, " Take, eat ; this is my body. Drink ye all of this my blood which is shed for the many."

Our life is but the surface embroidery worked upon the strong warp and woof of other men's services, and of dead men's deeds. We have taken of the life of every soldier of history who fought for the right and the liberty that we enjoy ; we have taken of the life of every student and investigator who wore out life and mind for us ; of every legislator and judge who by his own self-renunciation kept pure for us the ideas and practice of justice. In the smallest details of life the law is also absolute ; the coal with which we warm ourselves is ours because men parted with some of their life—perhaps their whole life—to dig it for us ; a spiritual eye can see the bones of smothered and buried miners glowing deep among the burning coals of our hearth-fires.

The unity of all life, that of past, present, and future, being bound by the laws of causation and heredity into an indivisible organism, makes duty to humanity, in a biological and literal sense, inevitable. Moreover, there is a cumulative action both to all sin and to all goodness, that especially makes us the present responsible focus of the struggle of the whole past, and the carriers to the future of our inheritance. Thus nothing is truer than the vicariousness of all lives, and our consequent obligation to worthily hand on the torch of life, formed out of and burning with the sufferings and sacrifices and heroisms of all past men and women, and not selfishly to use it for our own individual lighting, nor to burn it out in the dark ways of private sin.

It is the recognition of the unity of the present and future life, and the delight there is in participant work with God, that give the saint, the hero, and the martyr the courage to do their deeds and to live their lives. In choosing as they do they know the suffering that will inevitably come to them—the poverty, the non-success, and the malignity of the selfish. But they are willing to work for the future because to them the future is their own quite as much as the present is theirs. Modern heroism consists in daring to be poor, in quiet self-renouncing labor to thwart and counteract the wrongs of the recklessly selfish, and to help the Father of Life to transform the evils of the present into the large goodness of the future. It is not in the least a fantastic or unrealizable ideal to renounce selfishness and to work for the better future that shall rise after these our centuries have passed. The religious philosopher recognizes that his own individuality must be enlarged into divinity, and that in patient right-loving work for coming generations, he is working for self,—because self in its last and best analysis is God. Suffering and renunciation are not felt because of the anesthesia of duty, because of the enthusiasm of the friendship of God that is vouchsafed to them, because of the sweetness of the divine Being of which they are sharers, and because of the joy of the divine task, to which as co-workers they are called. The pain of personal tragedy is lost in the smile of peace, and, with the steel in one's heart, the lips calmly murmur, *non dolet, Pater, non dolet!*

Men's lives themselves are ratable in money values. Our greatest capitalist could but lately have gone to Brazil and could have bought outright something less than half a million of enslaved human beings. In the labor-market to-day the average wage in our country is about $300 a year, or an average life is worth about $6,000.

It is plain that the extreme of wasteful luxury of our rich citizens is an appalling sin. Without lessening the stimulus of enterprise due to desired wealth, society must legally demand accountability of its criminal rich, and must undertake to limit the acquirement of great fortunes. A sharp inheritance-tax is not only not unjust, but is better for the heir himself. Fortunes drawn directly from the labor and suffering of the many, by the Machiavellian cunning of interested law-makers, are such outrages that they should not be longer encouraged. The toadying to the rich by the poor is one of the incentives of wealth-hoarding, and when an enlightened general public shall by social fashions and acts condemn the parasitic profligate, wealth used solely for self-gratification will make the user so ashamed that he will join the ranks of the good and those of simpler lives. The infamy of a rich *roué* pretending to be a Christian, and of millionaires worshipping a God who condemned them and their riches as Jesus did, must soon become too ludicrous even for themselves and their worshipful priests to stomach. The sinfulness of mis-acquired and misused wealth arises from the fact that the progress of the systematization and security of the nutritional problem in civilization is thereby hindered and millions are chained to the slavery of insecurity. If all crops should fail, the world would starve to death in about two years. That in some aspects and ways wealth serves to increase the security, and that release of millions from that slavery would frequently result in the misuse of the freedom, is no valid argument against the essential principle. There are many things that money cannot buy, but which one cannot obtain without some money. So fast as men become capable of using leisure for God's purposes, so fast will the fact and misuse of wealth be limited. So long, therefore, as the poor misuse their little leisure, they have no argument to bear against the rich for misusing their great leisure.

There is thus always a return to the truth of the spiritualization of power and character, as being Life's end and aim. The good man, whether rich or poor, will be he who uses time, opportunity, and power to help men carry out the extension and perfection of Life. The selfish rich man's sin is greater than that of the selfish poor man, because his opportunity and power are greater ; and, as the immortal author of *Unto This Last* has incontrovertibly proved, a rich good man is quite a contradiction in terms. If he is good, he will soon become poor by aiding the aspiring poorer to realize their hidden lives, and in helping on God's evident purposes. The poor seek relief from labor, but labor, even some physical labor, is the condition of health both physiological and moral. It is excess of labor, impossibility of rest, and labor in deadly conditions and circumstances, that are the real evils.

The care, guidance, and best use of self has peculiar value from the fact that our body, mind, and life are the powers and things in this world over which we have the most complete control. Individualization is God's method of dividing work, and of deputing responsibility. It follows that an object and a reality lie behind the process that are far greater and more valuable than any part of it, and thus arises the logical need of modesty and of a constant guard against over-valuation of self. In making us a deputy, and in setting apart for us a limited field of work, the autonomy of self is created; but even this government is a limited monarchy, never an irresponsible autocracy. A ruler who uses his power with unselfish regard for the good of his people will be given even unconstitutional powers, and so, as we learn the right use of this wondrous mechanism of self, the limitations of our control and power with it are by God most willingly widened. This He is glad to accord, and only men's disloyalty and selfishness prevent a marvellous extension to them of power and freedom. To every one are given some peculiarity and uniqueness of capacity or of responsibility. The chief duty to self therefore consists in utilizing the gift for the benefit of Life's purposes.

The failure to thus improve the talent entrusted to us comes from many causes; for example: slavery to toil, or the great necessity of nutrition. This may often be lessened by wise renunciation of luxury, and by the enjoyment and utilization of those simple things that at last constitute the great good of life. Most people, even the very poor, waste sustenance and leisure. To this end a careful cultivation of the devising intellect, such as God shows in every organism, would largely spare the pain and friction of life. Man is almost the only wasteful animal in the world. Another method of utilization of life lies in the forethought and in the heroic energy to realize the essential work of life without sad experimentation and preliminary failures. Toward the end of life we all see how we could have come to our life-work quicker and more certainly had we but followed the more or less clearly seen way thereto, instead of acting upon whim and immature resolutions. We allow our young to blunder and stumble into life, which is as bad as the former method of driving them from birth into a cast-iron machine of circumstance by parental authority or class-condition. Cannot we learn the trend of character in young minds, and by kindly sympathy and wise direction get them at their proper life-work earlier and better? Only few know how much the work of life and the elevation of civilization depends upon right education and adequate training. We greatly need a class of good books to put in the hands of young people, books that shall aid them to a healthy self-knowledge, and teach them how to make the most of themselves, how to decide what are worthy aims and what is true success, and to keep them from wasting their lives in futile attempts, self-delusions, and ignoble desires.* It is the country lads, the farmer's boys and girls, that need these most; those far removed from the influence of brighter minds, and that are dumbly groping their way to an unseen light. These need to be taught to see the world about them, to understand it, and to utilize the wasted values and

---

* " Oh that I had known the art of life, or found some book or some man to tell me how to live, to take exercise, etc. But I found no one and so here I am."—THEODORE PARKER (*shortly before death*).

" It is not to die, or even to die of hunger, that makes a man wretched. Many men have died ; all men must die. But it is to live miserable, we know not why ; to work sore and yet gain nothing ; to be heart-worn, weary, yet isolated, unrelated, girt in with a cold universal Laissez-faire."

CARLYLE.

opportunities slipping through their hands and past their doors every day.

"This world is a world of joy to me, because I have tried to do my duty." These are the words of a noble man, and this is the experience of every normal-minded person. Tragedy and suffering may in part result from heroic adherence to duty, but they are more certainly due to other qualities and characteristics too rigidly and morbidly pursued. What we need to make us all happy is easy renunciation of worldly ambitions in obedience to serene and clear ideals of honor and duty. Dare to be poor, and unfamous! In all the world and in whatever position one finds himself there is joy to be had, and a peaceful brightness. Almost everywhere there are sunshine, trees, grass, breezes, health, and delight in being. And even where these are not, there is the possibility of a fountain of subjective happiness that would spontaneously bubble in lightsome laughter if we would but permit the playful divinity at the heart of us to do His will. Give the rein to God! Take your little tragedy out of God's way, and let Him flood your life with His rich love and ample content. Do not be so selfish, so earnest, so morbidly careful of yourself, and of the future. Let Him have His way. Observe Him, and learn of Him. Let the avaricious have the money, and let the ambitious have the fame; let us never give place in our own hearts to envy. Why should we not live in the day and bask in its brightness—why not love even those that are hateful and that are unjust to us? Our own happiness is too precious to spoil it with hatred, and the world's happiness is too easily lost to endanger it with our personal animosities. Nothing is more apparent to me than that the dear Father of Life is a calmly joyous and serenely happy

Being. The tragedy and suffering of the world's past have been the inobviable consequences of the struggle of Life to get Himself firmly entrenched in the inorganic world. The struggle is in large part now over. The domination of physical forces is assured, and henceforth the world is to blossom as a garden. Out from every wondrous bundle of material atoms reined by Biologos into a marvellous mechanism of organization—out from every swirling system of cells there shall gush the larksong of the visible-invisible, victoriously-happy God, singing because song and gladness are the very heart of Him. Oh, if we would only ourselves be happy; if we would be aidful and kind to others; if we would help Him,—nay, if we would only allow Him to be happy, Him in us!

What we all need is loyalty, firmly seized and devotedly adhered to, from childhood to the end; Loyalty to our aspirations, planted always in the young bosom by God's own hand; loyalty to purity, heaven's own light, that we bring with us into life, that "lies about us in our infancy," and that never need "fade into the light of common day"; loyalty to honor, that patrimony we too soon barter for less valuable things; loyalty to reason and intellect, that, if adhered to, would save us suffering and tragedy, self-deception and will-o-the-wisp swamps; loyalty to our true inner and higher self, which itself is literally loyalty to God. If, as believed, the diffused yet unitary being of God, while retaining true personality and self-centering, is the infinitely divided life and government of every living cell, and the ultimate reality or mind of every individualization, it follows that our duty to Him is clearly that of carrying out the work we are set to do, the work of fusing loyally our share of the labor, all our powers and influences, into the greater work of which we are, and are to perform, a part. The plainest fact of individualization and incarnation is divided function, deputized responsibility, a work to be done by infinite agencies, but all of them directly held of God. The human being is the highest and most trusted of these agencies, the one to which all others lead, and the divine Father is clearly awaiting the ripening process of incarnation and the recognition of their office by His highest children, in order to hand over to them progressively greater fulness of life and authority. To accept His work and His aims as our own is the essence of our duty to Him. In this loyal acceptance and righteous use we come to an ever-increasing sharing of the divine nature and personality, a likeness to and union with Him. In this way we love Him, as the infinitely lovable; sympathize with and help Him, the patient, ever active, and kind; revere Him as the holy and pure ideal to which we draw ourselves and are subtly drawn. But we are not abashed or overpowered by His presence, our faces are not solemn, nor are our prayers groans. Laughter, humor, and wit, at first hidden by the sternness and rigor of the struggle with matter in the beginning of the incarnation-process, break forth so soon as the serious work and sure foothold is secure, break forth ever more in man, the being nearest Himself. The brightness and gayety of children fresh from His hand are perpetual intimations that the divine Father of exuberant life delights in play, and is Himself a glad and happy being. Finally, the superb and elusive smile of beauty, gleaming through the evolution-tragedy and lighting the hills of all life, tells us that we are only at the beginning of the day and of a revelation of Himself by new attributes of spiritual joy and glory.

CHAPTER XIV.

BEAUTY.

THOUGH prevented from any extended discussion of esthetics, I cannot forbear mention of a few thoughts with which I have been impressed about the beautiful. Here is something that should palsy the tongue and pen of every atheist, materialist, utilitarian, or necessitarian. Its existence confutes all philosophies except those predicating a free and exquisite intelligence in Life. It is the most convincing of all "arguments" of God's existence, and of what is better than mere existence, of His non-utilitarian, non-solemn loveliness. It appeals with immediate power to all those who have eyes to see, or with logical power to conclude from perfect premises to certain conclusions. Even if "accidental" it is logically unaccountable to any except the God-perceiver. But it is not in any sense accidental. It is plainly introduced into the incarnation-process as a pure *ab extra* gratuity, an often expensive luxury and addendum to that process, thrust into it as if with conscious purpose to throw a light of hope and joy into the bitterness of the struggle, and to shoot among the darkness of our materiality a brightening shaft of heaven's own light.

I wish particularly to suggest notice of the fact that beauty is ready and waiting to endow all things whenever the overmastering necessity of nutrition does not forbid. Thus well-armed and stinging insects are usually endowed with high colors. Drummond says that two families of African butterflies are inedible, and these have brilliant colorings and are bold in showing themselves. The edible are neutral-colored and of hiding habits. The beautiful-plumaged bird is not usually the sitter, which is refused adornment in order while brooding not to invite attack. The covered nest is an expense of the exceptional bright-feathered sitter. Birds that can escape danger, as the lightning-like humming-bird, that superb incarnate glance of God's eye at the flower (as if some one said there is a single blossom not beautiful!), are at once robed in effulgent beauty.* Woman was evidently endowed with her splendid charm quite early enough to bring war and tragedy into all history. The flower is not a mere splash of color—which would have been sufficient to attract the bee's attention,—but a symmetry of tint, shading, and harmony, as if an angel's love-song had been crystallized into the folded and rainbowed petals, its perfume the breath of his sighing lips. A thousand

* "In their plumage, as Martin long ago wrote, nature has strained at every variety of effect and revelled in an infinity of modifications. How wonderful their garb is, with colors so varied, so intense, yet so seemingly evanescent!—the glittering mantle of powdered gold; the emerald green that changes to velvet black; ruby reds and luminous scarlets; dull bronze that brightens and burns like polished brass; and pale neutral tints that kindle to rose and lilac-colored flame. And to the glory of prismatic coloring are added feather decorations, such as the racket-plumes and downy ruffs of Spathura, the crest and frills of Lophornis, the sapphire gorget burning on the snow-white breast of Oreotrochilus, the fiery tail of Cometes, and, amongst grotesque forms, the long-pointed crest feathers, representing horns, and flowing white beard adorning the piebald goat-like face of Oxypogon. Excessive variation in this direction is checked in nearly all other birds by the need of a protective coloring, few kinds so greatly excelling in strength and activity as to be able to maintain their existence without it, etc."
HUDSON, *Of Humming Birds.*

examples rush to the mind. Moreover, and this note well, why the sensibility to beauty? Why does it delight and thrill us? No utility is served by it, and, if in scientific jargon, it is "the reaction of the organism to the environment," whence did it get into "the environment," and whence the percipient's pleasure? Only this alone can explain the outer or the inner fact, this, that as we are of God's own nature, and as He loves beauty, so there is the necessary charm out there thrilling us in here. He cannot hide Himself if He would, and even in the midst of the greatest of His biological difficulties and dangers, no leaf or hair or organism but is tell-tale of the artistic divinity. I have never seen a living thing that was simply and only useful. Like a lover's touch the esthetic caress lingers warmly and clingingly to the last, and is the one memory that does not fade.

And it is He that has given us those wondrous sense-organs which make beauty for us even out of the unliving world,—like unto the fairy anointing of human eyes that ever thereafter saw fairy-land in the most vulgar of prosaic things. It is the eye that creates color and light, the ear that makes sound and music. The rainbow and sunset are not out there, as we well know, but as we are always forgetting. Ethereal and aërial waves are *per se* most uninteresting, mechanical, and meaningless things. But these transforming and transmitting living mechanisms deliver to the mind products of unearthly beauty and supersensual significance. The ingenuity and love of the glad divine Artist in thus outfitting us with a responsiveness infinitely precious, various, and charming, to such insignificant, infinitesimal dead things as these oscillations of dead matter, call forth from our hearts a limitless gratitude and love.

What is the true and fundamental reason for the power and charm of beauty, for the slavery we all give to the beautiful woman, and for the pride she has in herself? Why is there an inexplainable glory in the brilliant and flashing eye, why a quivering sense of divinity in every inch of the Apollo Belvedere, or in living human flesh? Is it not simply the satisfaction of God Himself in His successful work? A million difficulties, fatalities, and unavoidable accidents have prevented success in other specimens, but, *Here*, says the divine Mechanic, *Here*, *behold success!* The successful ending of long human endeavor always provokes joy, elation, and satisfaction. God is just as glad to succeed as we are glad to succeed, and our delight in the charm of beauty is simply His delight—because we are He. Our ecstatic pleasure in beauty is the flush of the divine satisfaction at divine success. "And behold, it was very good!" Beauty, it must be remembered, is more than perfection: it is the smile of God overflooding perfection, as sunshine and swaying zephyr over hilltop or cornfield.

But after all, its greatest significance is that it is a promise and hint of what is to be. All young things, and especially children, are beautiful, because they are plastic and fresh from God's hand. It is apparent that God is trying to preserve this beauty into and through adult age, and who can doubt that He will by and by succeed. Then every woman will be a virtuous Helen, and every man a moral Goethe; perfect trees will laugh with the laughter of the skies, and all animal life will take up its line of march toward humanization, whilst we—"the eternal-womanly will ever lure us on."

CHAPTER XV.

SLEEP, DREAMING, AND AWAKENING.

SOME ancient philosopher, I think it was Plotinus, with deep feeling of the truth, said that "matter is the true river of Lethe; immersed in it the Soul forgets everything." Now although this expresses a certain forefeeling or partial aspect of the truth, it by no means gives any hint of the reason of the fact. The most obvious thought at once arising in the mind of the ordinary reader, the simplest objection to an acceptance of the incarnation-theory, is this of the loss of the divine self-consciousness upon entering matter in the incarnation-process, and also the non-recognition of the divine character and self as the essential basis of our own self-consciousness. This will prove no difficulty to those who will sympathetically ponder it, and with careful imagination conceive what must necessarily follow from the conditions antedating and synchronous with the incarnation-fact. It all becomes clearer if we resolutely hold fast the knowledge of these most obvious facts: that only by and through the cell-mechanism can God gain control of matter and mechanics; that individualization is the combination of cell-mechanisms for a specific purpose and function; that the higher or more complex cell-mechanism is dependent upon the ingestion of simpler ones—or higher forms of individualization depend upon the lower; that the control and function of cell-mechanism has never been given over to the individual, but has been retained entirely in the divine hands. In other words, the control of the essential mechanism of incarnation is by direct divine power, and the cell and all that pertains to cell-nutrition and function are mere dumb agencies of His clear grasp and intellectual use as instruments, subject to the help of an automatonization or regularity of action that insures a desirable relative uniformity of morphology, whilst it also aids in the systematization or outworking of the process.

So long, therefore, as to us is not given the control of cellular organo-genetic and nutritive processes, so long necessarily must the purely intellectual perception of God and of our divine nature be more or less defective. When physiology and thermo-chemistry have solved the mystery of cytology, the mind will begin to perceive God intellectually, because by that solved mystery will come the mathematical and mechanical demonstration of a supernatural and metaphysical agency. There is in the cell an as yet unestimated but certainly active foreign and non-mechanical force. The distance apart, and the difference of function, of the human and the divine mind are approximately estimated by this our ignorance of cell-life, but is more strikingly shown by the fact of our present inability to direct cytogenesis and cytotrophy—*i. e.*, cell-production and cell-nutrition. The individual consciousness is God's partial incarnation of Self, and must necessarily be nearly conterminous with the needs and functions of the organism as a whole. On the physical side cell-function has not been given to the individual control or consciousness, but only the direction of the combined results of cell-activities, such as molar motion, is thrown under human volition. On the divine side the human consciousness has been limited simply by the status

of the organism as regards comprehension of and loyalty
to the divine purpose. So fast as the individual has been
able to make right use of deputed responsibility, so fast is
it given him, and just so fast does the human consciousness enlarge into and become one with the divine. A far
greater bestowal of the divine attributes upon the individual
than is proportional to his educational and intellectual
condition, or to the grade of perfection attained, would
result in two obvious misfortunes: first, misuse of the
deputed power, from ignorance or disloyalty; and second,
unhappiness of the individual chained to a seemingly
ignoble task with a divine consciousness of greater capacities. I said "a far greater bestowal," because it is an
evident fact that in man knowledge, conscience, and
consciousness have outrun volition, possibility, and realization, and this has acted as a divine incentive to draw us
upward faster, to inspire us with a heavenly hunger and
with a heavenly hope. There would evidently be no
advantage to the incarnation-process, or to any individual
to outfit him with a divine consciousness immensely out-
running power and the duties of function. A grass blade
would have no more happiness or usefulness if given the
grade and degree of soul-life of a noble tree, nor would a
moth be the better off for having a lion's soul. If a
human mind were transferred to a horse's organization,
retaining the human self-consciousness, sensitiveness, and
knowledge, it would result both in wretchedness, and in
a very poor quality of equine function, at least until the
higher and useless attributes had atrophied. When the
parasitical *sacculina* fastens itself to the crab's tail, its eyes
and other sensory and motile organs become useless and
are soon lost. When the human being becomes a hermit,
or, in habits, an animal, the human soul also atrophies.

The progress of consciousness and the grade of soul-life are thus primarily conditioned upon (*not caused by!*) the grade of individuation, the degree of perfection of organization, and the necessary function. In the educated human being this individuation, or this subjection of the world by knowledge, science, and mechanical ingenuity, is fast reaching a comparative perfection that is denied to the individual as such. Civilization is merely a great tool or instrument, a new mental sense, and an indirect extension of body. Further progress of sense-making or of bodily mechanization becomes therefore unnecessary. The physiological (and with it the nutritional) problem is solved, and organofaction is at an end. No new physiological instruments will be developed, because none will be needed. All that is needed is the perfected use of those that have already been evolved, and the utilization of the indirect extension of the neurological and psychological instruments that we call science and civilization.

With a careful recognition of the singleness of aspect and limits of the metaphor, the three stages of the incarnation-process may be analogized by the facts of going to sleep, of dreaming, and of awakening, and epitomized by Froebel in the words, *From Life, Through Life, To Life.* In entering matter, God, in a sense, does enter a "river of Lethe," in so far as there is evidently that degree of task-setting, or allotment of the divine attributes required to carry on the necessary work of the process. When we drive the cows home we do not need or utilize our higher knowledge of trigonometry or of astronomy; and when we plant or train a grape-vine we do not use the higher knowledge of advanced pedagogics, or the conclusions and powers we have gained in sociological study and governmental administration. There is, indeed, a plenitude of the divine attributes, obvious or inferable, in the lower or lowest types of individuation. But it is most plain in cytology, or what is crudely called "the vegetative functions" of the organism, which, according to the metaphor, go on in sleep, and, as we should expect, go on even better in sleep than with the somewhat disturbing action of the waking personality. In the lower grades of the incarnation-process, God, so far as relates to the direct implication of the highest attributes of divinity, "goes to sleep," gives over to the cytologic mechanism (never, however, or chiefly an automatic or non-divine mechanism) the effectualization of His purpose, and reserves the higher consciousness to Himself.

A different sort and degree of the divine self-subdivision, or deputization of Himself, is that sharing of Himself that constitutes the mental or soul-life of the individual. The extent of this Self-giving, or the degree to which the higher attributes of God are grafted into the individual (plant, animal, or man), depends, as has been said, upon the perfection of organization, or upon the power to certainly control, the intellect to infallibly direct, and the loyalty to rightly use,—qualities that relatively develop *pari passu*.

The likeness to a dream of one's own life, the lives of all men, and of the course of cosmic life, has been recognized by the poets and philosophers of all ages. However like reality they were at the time, the objects we so earnestly and passionately struggled for in the past, as we now calmly smile at our old diaries, seem to be like the soft delusions of our dreams. All history seems like the phantom and aimless fancies of a slumbering dreamer. There is much nightmare in the historical dream, but there are also in it sweet figures, tender and sad, or lightsome and beautiful: St. Francis, Marcus Aurelius, Buddha, Beethoven, Heine, Goethe, one's mother, wife, or friend. And then there are the dreams of dreams: Helen of Troy, the Apollo Belvedere; Arthur, Elaine, and the Knights of the Table Round ; Dante's Beatrice, Siegfried, incorporeal and unrealized desire, and unseen yearned-for love, a hidden tragedy, a hunger for God. All we have done, and all that all men have done, seems like an aimless search for an unknown good, with the Maia-veil of illusion always withdrawing though never wholly withdrawn. Humanity is busied with its own brain-phantoms, delighted or frightened at the real-unreal visions spontaneously springing into the uneasy sleeper's imagination. There is a sense of unreality over and in all life, a subtle disturbing *Ahnung* that we are asleep, and that awakening is coming.

This sense of the dream-quality of life corresponds too truly to a philosophical knowledge of life to be passed indifferently by. All process, especially the incarnation-process, leads somewhither, is itself a means to an end, and to an earnest travelling soul the destination must be always more than the road of life or than the vehicle of the body. The dream-illusion of all living is the illusion of attaching interest only to the route, forgetting that we are definitely travelling somewhither and for an object. The roadway of life is rough or excellent ; the physiological carriage may be good or bad ; one's fellow-passengers may be pleasant or not; the scenery glorious or saddening; the journey long or short, comfortable or full of pain. Dreams are all these, however, and a distant church-spire or a chance philosophical pamphlet suggests to the dreamer that there is an approaching journey ending.

If the poor dreamer love the journey more than the home-coming, the dream better than the awakening, he will turn to thoughtlessness, to selfish plans and amusements, to the thousand distractions of life—war, money, or ambition—with which we poor travellers beguile the way and with which the way beguiles us. The traveller that finally turns with insatisfaction from all wayward distractions or amusements, reflects upon the outsetting of self and of all living things, of the meaning and object of the great journey of life, and of the coming return to the home;—such is the philosopher; and if in the little philosophic dream of this book we have dreamed duteously and wisely, the journey's end in death will not be a ruthless and sudden awakening among strangers, or to what we know not, but it will be the eager and delighted reception at home by our Father and our Friend. We have already, indeed, had the truer and better awakening, and death has nothing which we either fear or for which we hope.

The best that may be said of "the world" is that it is a letter direct from the Father's own hand, advising us, telling us of Himself, and urging us to hasten our return to Him. During the long journey we read it over and over again, delighted at the kindness it witnesses, and the beautiful suggestions it gives of His thoughtfulness and wisdom and lovableness. But, after all, the letter is only to recall to mind the living Being who wrote it, and it closes with the admonition to love Him and to come to Him. Blessed are we if the dear letter reach us long before our long journey's end, and thrice blessed we that receive such a letter from a Father that by the false message of a mistaken friend we had mourned as dead!

Out of the unreality and illusion of our dream-life there are two methods of awakening: the doubtful, mysterious, and unknown one that lies in and beyond death; the other, by the emotional, scientific, and philosophic perception of the significance of life, and our duty to the Father of Life. I urge, I fervently urge, that we need not wait for death to awaken us, but that here and now we may come to vivid self-consciousness. All these materiality-engendered dreams may be flashed out of the aroused mind by the divine call that we awaken, by the instant recognition of the meaning and the use of the whole biological process which scientific thought gives us, and that all religion has happily kept the mind open and waiting to receive. With the banishment of dream, we awaken—face to face with the smiling God.

INDEX.

A

Absorption of God in His work, 224, 225
Achromatism, 198
Adversity, uses of, 184, 185
Agriculture, 260
Altruism and egoism, 266
Anabolism and evolution, 148
Animals, anecdotes of, 172 ; care of, 264 ; intelligence of, 264
Animal-world, 261
Appropriation of God's mechanisms, 193
Architect, likeness of work of Biologos to, 201
Asceticism, 24, 164
Assassins, anecdote of, 230
Atoms, 14, 26, 151, 192, 216
Atrophic organs, 96, 168
Attention, 213, 215, 224
Automatonization of cell-function, 94, 95, 97
Avatars, 22
Awakening from dream of life, 291
Awe, 211

B

Badness of men, 182
Bats, blinding of, etc., 107
Beauty, 223, 279, 280, 283; and nutrition, 280 ; of woman, 163, 164, 170
Beethoven, quotation from, 112
Being and doing, 240
Biologos, attributes of, 16, 34, 213, 214 ; definition of, 16 ; purpose of, 248
Bird and feather ornaments, 263

Birds, songs of, 114
Blindness, 107, 109
Blood covenant and bloodshed, 124, 128
Blue color, 125
Booby's proof, 162
Books for the young, 276
Brahminism, 33, 35
Bridgman, Laura, 109
Brown-Séquard, reference to, 162
Buddhism, 32, 34, 35, 164

C

Calvinism, 199
Carlyle, quotation from, 276
Causes, ultimate, 4, 22, 23, 190
Cell-doctrine, 5, 61, 78, 99, 153
Cell, God's control of, 6, 67, 80, 84; nature of, etc., 5, 61, 67, 68, 78, 81, 99, 142, 158
Chakar, flight of, 10
Childlessness, 171, 172
Childraising, 234
Children, 163, 171
Christ, and Christianity, 38, 39 ; and his followers, 38, 51
Christian feeling, 50
Christianity, 36, 174 ; and civilization, 174, 175
Christ's message, 39, 43, 44
Chronicity of evolution process, 150, 153
Chronophotograph, 108
Civilization, 287
Cleanliness, 180
Clinging to life, 152
Clock, comparison of cell to, 160
Cold and heat, 106
Color-sense, 130

Colors, origin of, etc., 121, 124, 126, 127, 130
Color, symbolism, 131
Commercialism, 197
Conscience, significance of, 266, 267
Consciousness, individual, 285, 286
Contradictions of law of life, 248
Corbett and Sullivan, reference to, 197
Cosmic horror, 7, 189
Crowd-diseases, 179
Cruelty, to animals, 242; to trees, 243
Cytology, 77, 155; is medicine, 86, 179; is theology, 77
Cytolysis, 88

D

Darwinism, 15, 240
Darwin's theory of pangenesis, 99
Deaf-mutes, 107, 108
Death, 168, 178, 182; fear of, 235
Dedication, 171
Deductive truths, 23, 176
Defective classes, 163
Deforestation, 257
Delegation of freedom, 207
Density, value according to, 217
Design in inorganic world, 3
Determinism, 199; and evolution, 148; and materialism, 239
De Vries' theory, 98
Diseases, of civilization, 181; of plants, 258
Divinization of human, 253
Dog, anecdote of, 172
Doré's neophyte, 189
Dream, likeness of life to, 288
Drummond, quotation from, 281
Dualism of life and matter, 29, 30
Duties to world, groups of, 253
Duty and joy, 277

E

Eleventh commandment, 242
Endogamy, 158
Environment and biological process, 201
Ether, 26, 27, 216
Ether-waves, 116, 126
Ethical systems, faults of, 241

Ethics, 239, 243; systems of, 241, 251
Evil, 176, 185
Evolution, 145
Existence of God, proofs of, 51, 54
Exogamy, 158
Eye, functions of, 135, 136; history and development of, 138; influence of disease of, 138; reflexes of, 137; and sexual beauty, 137

F

Faith, the logic of, 151
Famines, 181
Farmer, the, 260
Fate, 3, 30
Fire, 122
Flowers, 281
Forestry, 74, 260
Freedom, 199, 227, 252; in animals, 203-205

G

Game of life, 268
Gamete, 158
Gilbert à Becket, anecdote of, 10
Gladstone, reference to, 127
God, character of, 177, 178; finiteness of, 5, 19, 20, 29, 31, 60, 177; in biology, 9, 59, 81, 84, 162; ingenuity of, 222; justice of, 177; kindness of, 162, 163, 177, 178, 189, 221; as substance, 216
Goethe, 36
Golden light, 123, 128
Goodness of God, 18
Gravity, law of, 4

H

Healing of wounds, 91, 93
Hearing, 112
Heat and cold, 106; of body, 88
Heroism, 272
Homesickness, 8
Homing instinct, 134
Hook-and-bait theory, 164, 169
Hudson, quotation from, 10, 114, 115, 261
Humanity, rights of, 255

INDEX. 295

Humming-birds, 281
Hunger, rôle of, 157

I

Idealism, 24, 140
Imitation of God, 24, 250
Immortality, 229 ; lack of evidence of, 235
Impiety, 19, 30
Incarnation, 58–65, 191, 219 ; process, justification of, 188, 196.
Indirect agencies of God's control, 246
Individuality, preservation of, 237
Individuation, 202
Infinitesimally small, God and the, 216
Infinity of God, 16, 20, 21
Ingenuity of God, 222
Inorganic world, duty to, 254
Intellectual perception of God, 285
Intelligence behind living things, 3
Inventions and imitations, 193

J

Joy and duty, 277

K

Keller, Helen, 109
Key of mysteries, 93
Kingdom of Heaven on earth, 246
Kitten, anecdote of, 172

L

Laughter and humor, 279
Law, nature of, 34
Lethe, river of, 284, 287
Life, aim of, 5, 190, 192 ; nature of, 14, 15, 19, 60 ; origin of, 4, 21, 22 ; work of, 23, 34, 69, 191
Light, 121, 123
Light-year, 13
Limitations and supplementations of eleventh commandment, 46
Living and lifeless things, distinction between, 3
Love, 157, 163, 164; anecdote of, 222
Loyalty, need of, 278
Luxury, 40, 206, 273

M

Magnus, reference to, 127
Marsh's *The Earth*, allusion to, 74
Material and worker, 12
Materialism, 24, 26, 28, 154
Matter and life, 12, 22, 28, 29, 47, 67, 78, 81, 186, 191
Matter, origin of, 4, 23, 31
Mechanicalization of cell-function, 94
Mechanics of incarnation, 65, 67, 104, 155, 159, 219
Medicine, 86, 92
Metaphysical, 14, 155
Mind, attitude of, as to mystery, 2
Molecule, 66, 67
Money, 271, 273 ; love of, 209, 210
Monism, 28
Morality and theism, 35
Moral law, 7, 8
Moses, order of, 259
Mother-love and motherhood, 171, 172
Music, 112, 113, 139, 197
Mystery, incuriosity concerning, 1, 23

N

Nature and man, 74, 75
Nightmare of the dream of history, 288
Nutrition, 85, 103
Nutritional difficulty, 150, 157, 158, 160, 161, 169, 179, 181, 195, 202, 214, 215

O

Omne vivum ex vivo, 153
Omnipotence of God, 16, 29, 47, 48
Omnis cellula e cellula, 155
Omniscience of God, 17
Organic crystallization, 156
Organic life, unity of, 69, 73
Organofaction, 94, 202
Origin of inorganic universe, 3

P

Paderewsky, 197
Pangenesis theory, 99
Pantheism, 26, 28, 61

Parker, quotation from, 276
Park, Professor, anecdote of, 115
Patents, 255
Pathogeny, 179
Personality, 211, 221, 224; origin of, 223; true, 236
Pessimism, 35, 164, 182; truth of, 267, 268
Philosophic dream, 290
Philosophy and religion, 10
Physical and metaphysical, 12, 154, 196
Physical world, 13, 25, 186
Physician, the, 197
Piety, 19, 212
Pineal eye, 134
Plant-life, 71, 72, 124, 129
Plant-world, use of, 257
Plotinus, quotation from, 284
Politics, 197
Poverty, 184; of wealth, 184
Power, use of, over others, 265
Pressure-sense, 105
Prevention of cruelty to animals, 242
Progress, 223
Proofs of existence of God, 51
Prosperity, 184
Purpose of God, 190, 191
Pyrolatry, 122

R

Red color, 124, 127, 128
Rejuvenescence, 150, 157
Religion, 9, 40, 42, 44, 45; of science, 49; of the future, 174, 175
Renunciation, philosophy of, 268; of divine attributes, 253
Reproduction, 85, 157, 165, 173; of lost parts, 100
Responsibility, delegation of, 207
Retina, sensibility of, 120, 127
Riches, limitation of, 273
Riddle of life, 3, 93
Rights of humanity, 255
Rule of ethics, 243, 245, 250

S

Sagacity and morality of plants, 256
Scaffold and steeple of life, 237
Schopenhauer, 36, 164, 188, 199

Science, 9, 82; a failure of, 151
Self, duty to, 275
Selfishness and unselfishness, 270
Self-love, mechanisms of, 266
Senescence, 150, 158, 161
Sensation, 101, 110, 140, 142
Sense-making, 287
Sense-organs, 282
Sex and sexualism, 157, 164, 165
Sin, essence of, 269
Sleep, dreaming, and awakening, 284
Smell, sense of, 111
Solar system, 66, 67
Somacule, 67, 218
Sons of God, 226
Space, comprehension of, 211, 212
Spectrum, divisions or rays of, 118, 121
Spencer, theory of, 154
Spiritists, evidence of, 236
Sport, 261
Standard of conduct, 250
Star-lit heaven, 7–8
Stars, motion of, 4
Success, 206
Suffering, exaggeration of, 183
Suicide, 188
Sunlight, composition of, 121

T

Temperature, and its significance, 87, 89, 90; of plants, 259
Theism and morality, 35
Theology, 77
Time, comprehension of, 213
Touch, 105
Tragedy, of the spirit, 9; the divine, 224
Transmigration, 33
Trees, 256, 259, 260
Trumbull, reference to, 124
Truth, best method of reaching, 2
Tuberculosis, 180

V

Vegetarianism, 261
Vegetation, world of, 71, 72, 124, 129, 204; duty to, 256, 257
Vegetative functions, 288
Venereal disease, 180

Vibrations, the stimuli of sensation, 105
Vicariousness of lives, 271, 272
Victory over death, 235
Vision, 117
Visual opera, 139
Vivisection, 264
Voice, human, 115

W

Wagner, influence of, 139
Wallace, quotation from, 240
War, 125
Wastefulness of man, 275
Wave-length, 126
Waves, ether, 116, 126, 216, 217, 219

Wealth, 184; use of, 274
Weissmann's theory, 99, 158
Wilson, Dr., observations of, 259
Woman, beauty of, 281
Wood, use of, 258
World, likened to a letter from God, 290

Y

Yeast and muscle, 256
Young, books for the, 276

Z

Zeno's paradox, 154
Zero, absolute, 27

www.ingramcontent.com/pod-product-compliance
Lightning Source LLC
Chambersburg PA
CBHW021958220426
43663CB00007B/874